MATERIALS SCIENCE AND T

DYNAMIC PROPERTIES OF CONCRETE UNDER MULTI-AXIAL LOADING

MATERIALS SCIENCE AND TECHNOLOGIES

Additional books in this series can be found on Nova's website under the Series tab.

Additional E-books in this series can be found on Nova's website under the E-book tab.

MATERIALS SCIENCE AND TECHNOLOGIES

DYNAMIC PROPERTIES OF CONCRETE UNDER MULTI-AXIAL LOADING

DONGMING YAN
GAO LIN
AND
GENDA CHEN

Nova Science Publishers, Inc.
New York

For permission to use material from this book please contact us:
Telephone 631-231-7269; Fax 631-231-8175
Web Site: http://www.novapublishers.com

NOTICE TO THE READER

Additional color graphics may be available in the e-book version of this book.

LIBRARY OF CONGRESS CATALOGING-IN-PUBLICATION DATA

Yan, Dongming.
Dynamic properties of concrete under multi-axial loading / authors,
Dongming Yan, Gao Lin, Genda Chen.
p. cm.
Includes bibliographical references and index.
ISBN 978-1-61728-907-1 (softcover)
1. Concrete--Plastic properties. 2. Structural dynamics. I. Lin, Gao,
1978- II. Chen, Genda. III. Title.
TA440.Y36 2010
624.1'834--dc22

2010027132

Published by Nova Science Publishers, Inc. † New York

CONTENTS

PREFACE

Large-scale concrete structures, such as long-span bridges, hydraulic dams, and high-rise buildings have increasingly been exposed to multi-hazard environments in recent years. Understanding the behavior of concrete under earthquake and blast loads becomes more urgent and important in the design of these engineering structures. This book briefly introduces the research significance and measurementfor the dynamic properties of concrete at high strain rates. The mechanical parameters of concrete under uniaxial dynamic loading conditions were obtained and compared with the static properties that have been widely used in design codes. A special test apparatus and procedure was designed and presented in order to evaluate the main concrete characteristics under multiaxial dynamic loading and understand various failure mechanisms in one-, two-, and three-dimensional loading conditions. The failure modes of concrete were investigated. Based on a comprehensive test program, material specifications are recommended for the seismic design of concrete structures.

Chapter 1

INTRODUCTION

Buildings and civil infrastructures have increasingly been exposed to multi-hazard environments in recent years. As earthquakes,collisions, and explosions occur more frequently, understanding the concrete behavior at high strain rates becomes more urgent and more important in the design of civil engineering structures. In these cases, the strain rates can be of different orders of magnitude as shown in Figure 1. Most engineering materials exhibit load-rate or strain-rate dependent properties such as strength and deformation. For concrete materials, the strain-rate effect is especially important in the design and analysis of large-scale structures such as arch dams, nuclear reactors, and bridges.

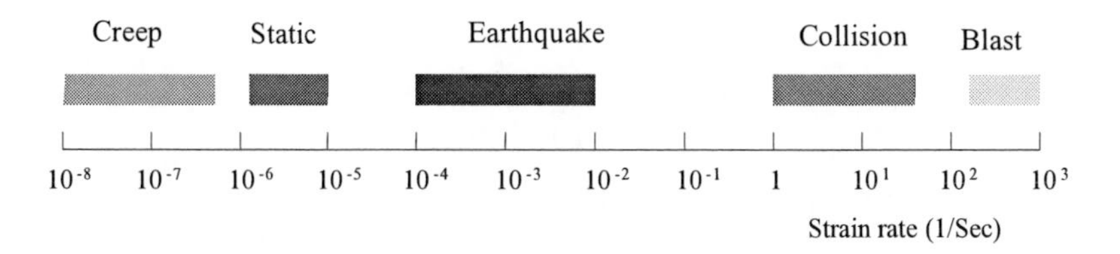

Figure 1. Magnitude of strain rates expected for different loading cases.

The rate-dependent behaviorof engineering materials is affected by environment factors to a great extent. These factors, such as temperature and moisture content, have to be taken into consideration, because concrete may be dry, partially wet or completely wet. In some areas of China, in winter concrete may be exposed to a temperature as low as -30°C. Understanding the dynamic behavior of concrete under various circumstances is an issue of great significance for application in civil engineering.

The early investigation on the effect of strainrate on concretepropertiesdated back in 1917 when Abrams observed the rate sensitive behavior of concrete during compressive tests (Adam, 1985). Since then, a number of other studies have been carried out as indicated in Figure 2 and Figure 3to understand the dynamic properties of concrete (Bischoff and Perry, 1991; Malvar and Ross, 1998; Rossi et al., 1996; Mellinger and Birkimer, 1966; Ross et al., 1996). This is especially evident under compressive loads due to ease in measurement.

Various test devices and procedures have been used by many investigators (Malvar and Ross, 1998), including impact of a dropping weight, splitting Hopkinson pressure bar (SHPB) and close-in explosions. The size of the specimens used in these studies is relatively small, for example, 2 in (51 mm) in diameter for impact tests (Mellinger and Birkimer, 1966) and 0.5 to 2 in (12.7 to 51 mm) in diameter for SHPB (Ross et al., 1996; Tedesco et al., 1993). Most of the tests were conducted in uniaxial stress state, either compressive or tensile.The effect of strain rate on the strength of concrete is typically represented by a dynamic increase factor (***DIF***), i.e., the ratio of dynamic to static strength versus strain rate on a semi-log or log-log scale.

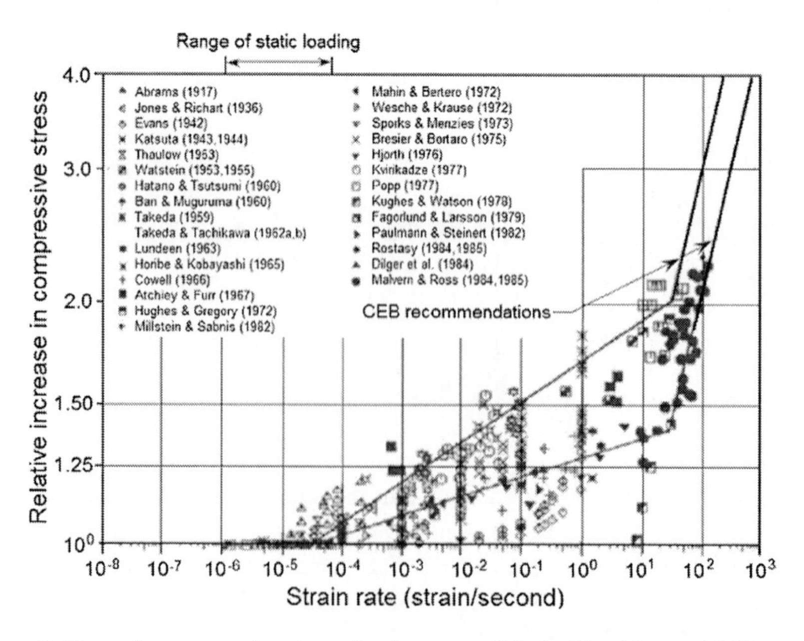

Figure 2. Dynamic compressive strength of concrete (Bischoff and Perry, 1991).

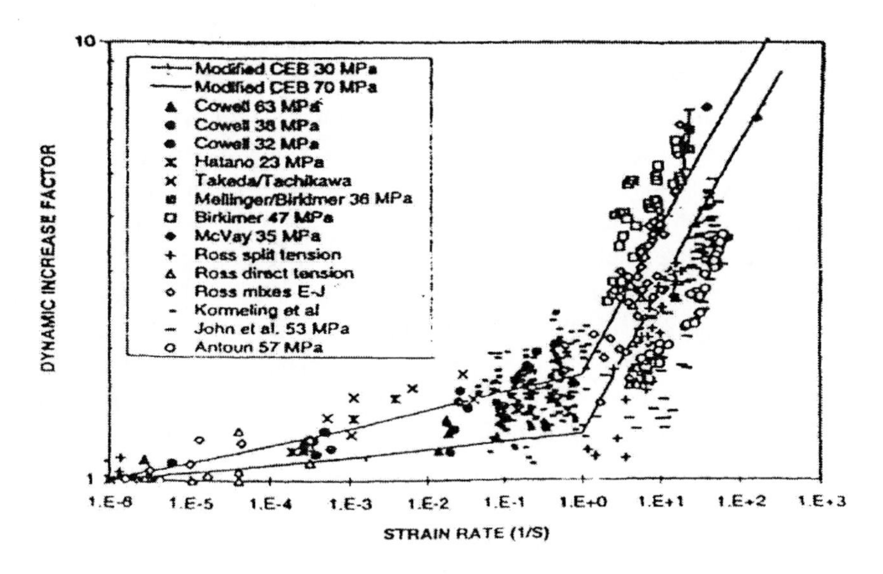

Figure 3. Dynamic tensile strength of concrete (Malver and Ross, 1998).

Although many concrete structures such as hydraulic dams are subjectedto various states of complex stresses, the dynamic behavior of concrete in multi-axial stress state isnot well understood due mainly to lack of proper test facilities. Currently, the available performance criteria and constitutive laws of concrete under high strain rates are almost all inferred from the experimental data of uniaxial tests; they have not been verified with the dynamic test data of concrete in multiaxial stress state. For example, Bicanic and Zienkiewicz (1983) extrapolated the dynamic behavior of concrete from uniaxial test results and established the dynamic constitutive law of concrete. Adam (2003) introduced a dynamic strength criterion to account for the strain-rate effect on concrete properties. The two constants in the strength criterion equation were approximately determined from uniaxial compression tests. However, stress state significantly influences the dynamic behavior of concrete. According to Gran et al. (1989), for Hopkinson bar specimens made of high strength concrete, the shear failure envelope at strain rates between 1.3 sec^{-1} and 5 sec^{-1} was 30% to 40% higher than that under static loading.

Limited studies of concrete in multiaxial stress state have been conducted to understand the dynamic behavior of concrete in practical applications. Takeda and Tachikawa (1962) studied the dynamic failure criteria of concrete with triaxial tests conducted under a constant confining pressure and at an axial strain rate of $(0.2\sim2)\times10^{-5}$ sec^{-1}, $(0.2\sim2)\times10^{-2}$ sec^{-1} or $(0.2\sim2)\times10$ sec^{-1}. It was observed that the concrete strength was considerably affected by strain rate. Fujikake et al. (2000) tested cylindrical concrete specimens using a triaxial loading apparatus at strain rates ranging from 1.2×10^{-5} sec^{-1} to 1.2 sec^{-1}. It was concluded that, although it had little influence on the failure mode of the tested specimens, the strain rate affected the maximum strength of concrete in triaxial stress state. The concrete strength increased with the strain rate but the increment in strength decreased with the confining pressure. Ahmad (2003) reported the results of an experimental investigation of hoop-confined concrete at strain rates of 3.2×10^{-5} sec^{-1} and 3×10^{-2} sec^{-1}, and concluded that the secant modulus of elasticity, the ultimate strength and the ultimate strain were increased at high strain rates for both plain concrete and hoop-confined concrete. It is noted that the confinement pressure in these experiments varied with the applied load due to the increased strain in hoop reinforcements.

This book aims to introduce the dynamic characterization of concrete in uniaxial, biaxial and multiaxial stress states for various applicationsto earthquake engineering. It is organized with six chapters. Chapter 1 introduces the significance of strain rates to understand material properties. Chapter 2 overviews various test methods for dynamic properties of concrete that are currently available in the research community, including the use of a unique test apparatus designed by Dalian University of Technology, China. Chapter 3 presents the uniaxial dynamic behaviorand fracture mechanism of concrete at high strain rates. Chapters 4 and 5 introduce the mechanical properties of concrete under biaxial and triaxial stress states, respectively. Finally, Chapter 6 summarizes the main findings from this study with design recommendations for structures in seismic regions.

Chapter 2

TEST METHODS IN CONCRETE MATERIAL DYNAMICS

Various test methods have been employed to study the dynamic properties of concrete (Bischoff and Perry, 1991). Depending on the expected strain rate range, experiment equipment can be generally classified as servo-hydraulic loading machines, drop-weight or falling weight equipment and SHPB testing setup.

In general, hydraulic test machinesare used to conduct static test on plain concrete or cement at a static strain rate from 10^{-7} to 10^{-3} sec^{-1}. However, the pumps and valves of testing machines can be specially designed to facilitate the accelerated flow of oils and build up the required pressure in a short time, reaching a strain rate up to 1 sec^{-1}.

Drop weights have been commonly applied in dynamic experiments over a strain rate of 1 to 10 sec^{-1}. The height and weight used in this test method have to be determined in advance. As the weight starts falling, the actual loading condition of a test specimen is difficult to control.Figure 4shows a drop weight system adopted by Elfahal et al (2005).

Hopkinson Splitting Bar (HSPB) method has been widely used to investigate the behavior of concrete at high strain rates up to 10^{2} sec^{-1}(Figure 5). By using a gas gun, the strain rate can be increased to above 10^{2} sec^{-1}. Generally, the SHPB method has widely been used to study the behavior of concrete at strain rate beyond 1.0 sec^{-1}.

Since the main objective of this book is to study the dynamic behavior of concrete subject to seismic loading with a strain rate of 10^{-4} to 10^{-2} sec^{-1}, only the hydraulic testing machines are introduced in this section. Specifically, the uniaxial tensile properties of concrete was obtained with anMTS testing machine while the uniaxial, biaxial and triaxial compressive behaviors of concrete wereinvestigated with a customized multiaxial testing machine designed and manufactured in the authors' laboratory. In order to understand the concrete behavior,the procedures for obtaining the dynamic properties of concrete are addressed as well.

Figure 4. Falling weight test system (Elfahal et al., 2005).

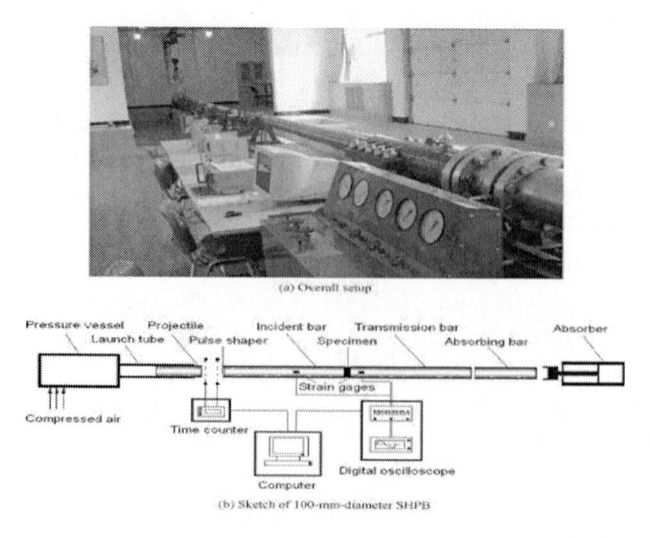

Figure 5. SHPB apparatus (Li and Xu, 2009).

DYNAMIC BEHAVIOR OF CONCRETE AT HIGH LOAD RATES: TENSION TESTS

Specimens

Concrete Mix Design

This bookdeals with the mechanical properties of concrete under uniaxial, biaxial and triaxial stress states. Limited by the testingcapabilityfor dynamic experiments in triaxial stress states, concrete materials of low compressive strengthwere used in various tests. Specifically, the mix designs of two types of concrete as specified in Table 1 includedthe type 42.5R Portland cement, river sand, crushed gravel with a maximum grain size of10 mm, and tap water. Their 28-day strengths are presented in Table 2. It was observed that the test data acquired from any specimen fell within a 15% difference from the average of its group. This ensured the reliability of the test results. It was assumed that the concrete specimens were isotropic before or in the testing.

Table 1. Concrete mix designs used for specimens

Concrete type	Cement	Water	Gravel	Sand
Mixture A	1.00	1.02	5.35	4.38
Mixture B	1.00	0.69	3.93	2.63

Table 2. The 28-day strength of concrete (MPa)

Concrete Mixture	Compressive Strength	Splitting Strength
A	10.7	0.83
B	21.2	2.33

Size and Curing Process

The specimens used for direct tension tests were in dog-bone shape, as shown in Figure 6. Compressive tests of concrete were tested on $100^{mm}\times100^{mm}\times 100^{mm}$ cubes. All specimens were cast in steel molds and compacted by a vibrating table. One day after the steel molds were removed, specimens were immersed in a water tank for another 2 days, cured in a fogroom for 28 days, and then air-dried in the laboratory. The average relative moisture content of concrete samples was6.3%.The specimens in each group were tested within 10 days to ensure the minimum change in their concrete properties during the period of testing.

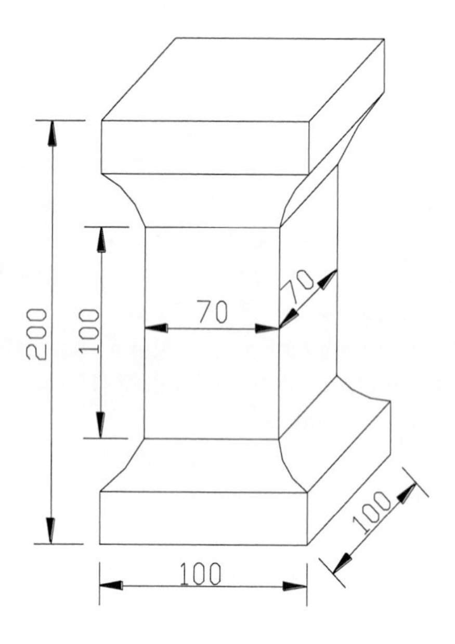

Figure 6. Dimensions of a dog-boneshaped specimen (unit: mm).

Test preparation and Condition

In this study, specimens of different strengths were tested undervariousmoisture and temperatureconditions. They were divided into five groups: HDR, LDR, HWR, HDC and HWC, wherein the first letter L or H denotes low or high strength concrete, the second letter D or W denotes dry or wet environment, and the third letter R and C stands for room temperature and cold temperature, respectively. Table 3 presents characteristics of the specimens in each group including water-cement ratio (w/c), moisture and temperature.

Each specimen was polished with sand papers to ensure all sides are smooth, scrubbing sundries out of the surfaces by acetone. It was then instrumented with a pair of crossed strain gauges on each side surface by epoxy to measure the longitudinal and transverse strains, respectively. Once the epoxy is cured for 12 hours, the strain gauges were covered with paraffin to ensure that their resistances are not affected by temperature and moisture. A frame was used to mount the two Linear Variable Differential Transformers (LVDTs) against two opposite sides of each specimen.

Table 3. Characteristics of the specimens

Group	w/c ratio	Moisture (%)	Temperature at testing (°C)
HDR	0.69	0.3	20
LDR	1.02	0.3	20
HWR	0.69	4.8	20
HDC	0.69	0.3*	-30
HWC	0.69	4.8*	-30

* Moisture was measured before specimens have been placed in an environmental chamber for 70 hours under cold conditions.

For direct tension tests, several methods to grip a specimen have been adopted by many researchers (Swaddiwudhipong et al., 2003). They include (1) rings on truncated cones (Elvery and Haroum, 1968); (2) embedded steel bars (Xie and Liu, 1989); (3) lateral gripping (Baishya et al., 1997); and (4)gluing steel plate (Yan and Lin, 2006). All four methods involve various degrees of loading eccentricity and nonuniformdistribution of stress and strain. The gluing steel plate method often experiences the poor adherence between the steel plate and moist concrete. The remaining three methods potentially contribute to stress concentration, causing fracture at the ends of a specimen. It is particularly a challenge to properly bond the embedded bars with concrete and ensure they are concentric with the specimen. For various test setups, the tensile stress distribution may differ significantly, as illustrated in Figure 7. Among the four gripping methods, gluing the dog-bone shaped specimen on a steel plate generally provides a more uniform stress distribution over the cross section of a test part of the specimen.

In this study, both ends of a dog-bone shaped specimen were attached to steel plates by JGN-II structural adhesive, which has a tensile strength of 35MPa. Each specimen in Group HDR and LDR was tested immediately after the surface preparation as described previously. Each specimen in Group HWR wastested after it has been submerged in tap water for 60 hours to ensure its full saturation condition. Note that submerging a specimen in tap water for a longer time may influence the hydration of concrete. Prior to testing, each specimen in Group HDC waskept in an approximately -30°C (-32 ~ -28°C) chamber for 70 hours. Each of the Group HWC specimens was covered with a layer of plastic sheet after being cured under water for 60 hours, and then put into the low-temperature chamber for 70 hours. In this way, only a small amount of water contained in the specimen was lost during the freezing process.

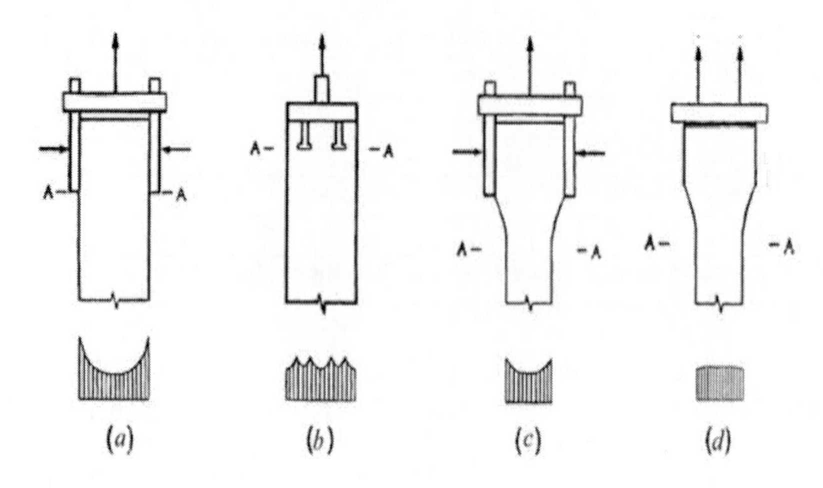

Figure 7. Stress distribution of specimens with different connections.

Testing Equipment

An MTS 810 testing machine with a programmable displacement-controlled loading system was employed to produce tensile loads at a maximum nominal rate up to 150mm/s. Its force capacity is 100kN and stiffness is 2.6×10^5 kN/s. The measurement system consists of a DPM-8H strain amplifier, an INV306D intelligent signal processor and DASP software. The machine can operate at the highest frequency of 10^4Hz. Foil strain gages, 50mm in length and 5mm in width, were used to measure strain histories during the dynamic fracture process. In addition, two pairs of LVDTs were fixed on each specimen to measure the displacement after visible cracks appeared. A tape recorder was also used to take data at high strain rates. Two computers were used for each experiment: one served ascontrol system and the other as data acquisition system (See Figure 8).

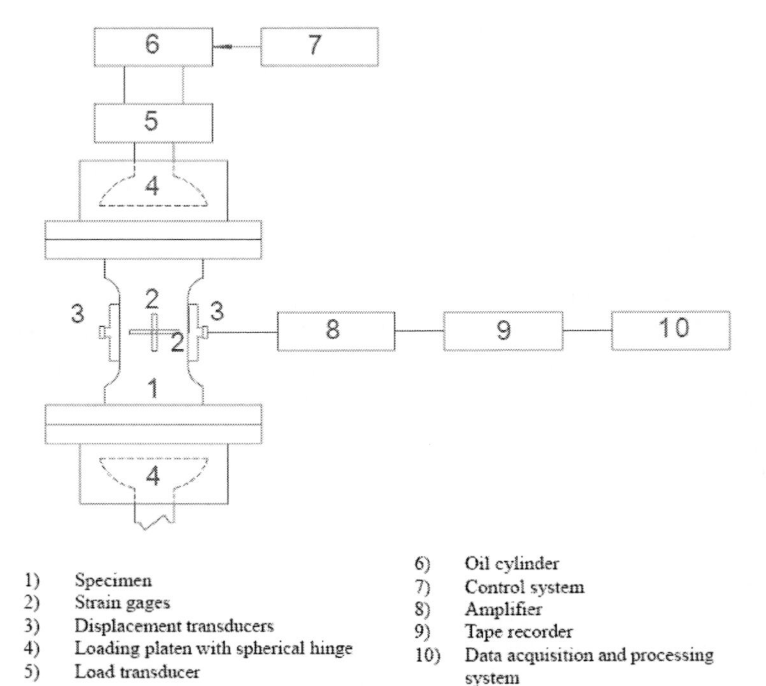

Figure 8. Schematic diagram of tensile testing system.

Test Procedure

The two end steel plates of each specimen were attached to the upper and lower heads of the MST testing machine with eight bolts in 18 mm in diameter each. A best effort wasinitially made to align the centerline of the specimen with the applied load before bolts were tightened up. The specimen was then subjected to an initial pressureof 0.5MPa to ensure that all four longitudinal strains have similar readings. If their difference exceeded a predetermined value, i.e. 8%,the specimen was unloaded and the bolts were loosened up so that the centerline of the specimen can be adjusted to its satisfactory position. Next, the LVDTs were deployed and adjusted for zero readings prior to loading. Finally, the specimen was loaded to failure. Each test was completed within 10 minutes.

Dynamic Behavior of of Concrete at High Load Rates: Compression Tests

Most of the existing models to account for confinement effects were developed based on the test results of reinforced concrete columns, using the estimated confinement and the assumed effective area of column cores. The column confinement provided by spirals, ties, steel tubes or other materials such as fiber reinforced polymers depends on the lateral dilation of the concrete under axial loads and the constitutive law of the confining materials. As a result, the models can only represent the approximate behavior of concrete under a multiaxial stress state. On the other hand, the servo-controlled hydraulic pressure system described above can directly provide a controllable confinement on concrete specimens.

Specimen

All specimens used for various compression testsare$100^{mm}\times100^{mm}\times100^{mm}$ cubes. The concretemix design used in this study was Type A as specified in Table 1. . Like tension tests, each specimen was surface prepared well prior to testing. The rough surfaces of the specimen were polished and any sundries on the surfaces were scrubbed out. Each of the loading faces was covered with three layers of plastic sheets and further lubricated with MoS2 grease in order to reduce the surface friction to a minimum. At least four specimens were tested for each test case to ensure the repeatability of experimental results.

Testing Equipment

To simulate the rapid dilation of concrete at a high load rate, the servo-controlled hydraulic pressure system must respond to the applied load in a short time. To this endeavor, a modern electronic servo-controlsystem was used to closely monitor and control the confining hydraulic pressure on each specimen in a multiaxial stress state.

Compression tests were conducted on the servo-hydraulic multiaxial testing system designed and built at DalianUniversity of Technology, China. The experimental apparatus is detailed in Figure 10. The testing system allows for free and independent motions in three directions. Along each direction, a pair of pressure levers loads a test specimen through two platens located on two opposite sides of the specimen. A spherical hinge is installed between a lever and a platen on the same side of the specimen to ensure that the load be exerted exactly along the load axis. The two pressure levers are connected with a load transducer and an oil cylinder, respectively. The nominal capacity of the testing system is 2000 kN in each direction.

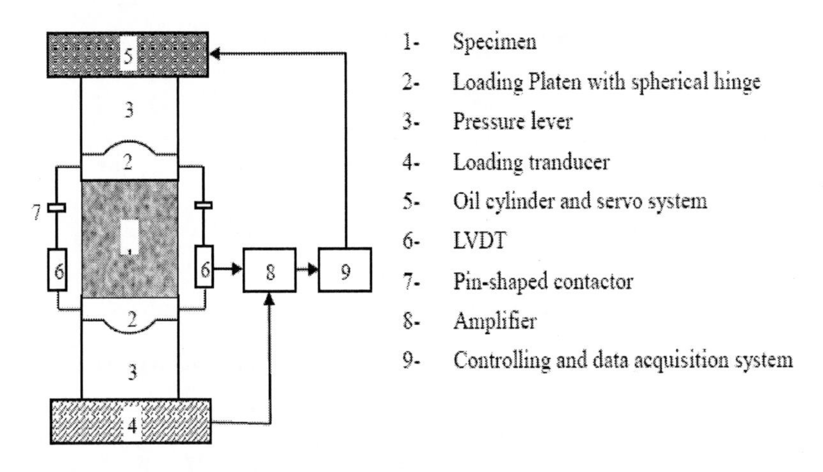

Figure 9. Schematic diagram of the triaxialtesting setup in one direction.

During testing, each specimen was instrumented with six LVDTs, two in each loading direction. Each LVDT had a stroke of 7 mm; it was attached to the two platens that were connected in series with the two opposite faces of a cubic specimen. The measured load and deformation were transmitted to the data acquisition and the processing unit of a computer through a specially allocated amplifier. They were then converted to stress and strain, respectively, using the undeformed area and length of the specimen. The complete stress-strain curve of the specimen including ascending and descending parts was recorded automatically. Displacement control mode was used to ensure the loading platen moving at a certain speed, which would lead to a relatively constant strain rate during testing. In the following discussion and analysis, concrete specimens were considered isotropic before and during tests.

Figure 10. Multiaxial testing equipment and apparatus.

To demonstrate the performance of the custom-made test device shown in Figure 10, both axial stress andaxial strain histories measured from two tested specimens are presented in Figure 11 when constant confining pressure was applied in two lateral directions and in Figure 12 when a confining pressure equal to the axial stress was applied in one lateral direction. It can be seen from Figure 11 that a nearly constant strain rate (low) and constant confinement have been realized during the demonstration test. Note that a slight fluctuation of the confining pressure over the time is due likely to the performance of the oil pump used during the test and the elasticity of mechanical components in the testing apparatus. Overall, the fluctuation level is within an acceptable limit (< 0.5 MPa). Figure 12 also indicates a nearly constant strain rate (high) and confining pressure that is equal to the axial stress. Note that a small change of the measured strain rate deems acceptable since any observable difference in material properties only becomes significant when the loading rate is increased by one order of magnitude or greater.Together, Figures 11 and 12 illustrate that the custom-made test device can realize a controllable strain rate (high or low) under various confinement. Note that the stress-strain curves of concrete under triaxial and biaxial stress states are quite different, signifying the importance of investigating concrete behaviors under multi-axial loading.

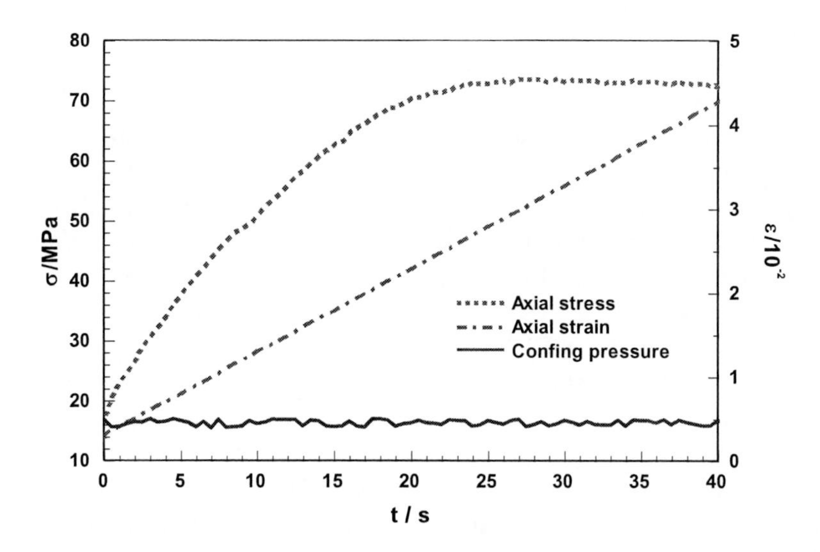

Figure 11. Measured stress and strain in triaxial conditionwith constant confinement.

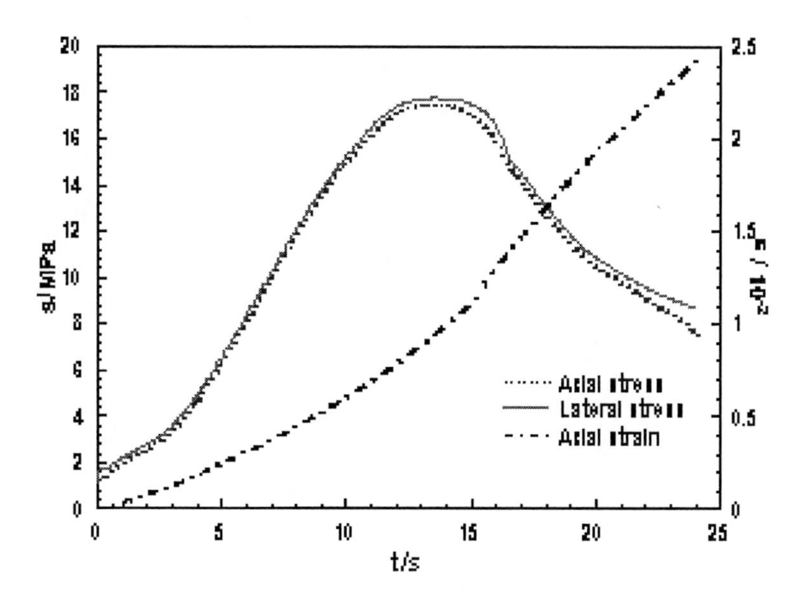

Figure 12. Measured stress and strain in biaxial condition with proportional confinement.

Test Procedure

Each specimen was first placed in the middle of the triaxial experimental apparatus. Three pairs of loading platens were aligned with any two opposite faces of the specimen and moved close but untouched to the specimen. Second, an initial pressure (1 MPa in this study) was applied in axial and two lateral directions on the specimen, which is automatically controlled by the in-house computer program. Third, lateral hydrostatic pressures were applied to the specimen step-by-step up to a predetermined lateral pressure at a rate of 0.001mm/sec. Fourth, six LVDTs were installed in three loading directions, respectively; and their initial measurements were taken and adjusted to zero in the computer acquisition system. Finally, the pressure load in axial direction continued to be increased at a constant rate until the concrete specimen crushed or experienced excessive deformation while the pressure loads in two lateral directions remained unchanged or proportional to the axial stress until failure of specimen. Four or more specimens were tested for each combination of strain rate and stress ratio to examine the repeatability of experimental results. The loading procedure was implemented automatically under the control of the computer program.

Chapter 3

CONCRETE UNDER UNIAXIAL DYNAMIC LOADING

In order to investigate the mechanical properties of concrete under earthquake loads, a strain rate range from $10^{-4}sec^{-1}$to $10^{-2}sec^{-1}$is typically interested according to Figure 1. Due to the capability of the MTS testing machine, a broader range of strain rates ($10^{-5}sec$, $10^{-4}sec^{-1}$, $10^{-3}sec^{-1}$, $10^{-2}sec^{-1}$, $10^{-1}sec^{-1}$, and $10^{-0.3}$ sec^{-1}) wasconsidered for specimens in Group HDR, LDR and HWR, and strain rates of $10^{-5}sec^{-1}$, $10^{-3}sec^{-1}$, and $10^{-1}sec^{-1}$were taken for specimens in Group HDC. The upper limit of the strain rate, $10^{-0.3}$ sec^{-1}, was controlled bythe capacity of the loading system. Considering the fact that the tensile strength of fully saturated concrete under low temperature -30°C is several times that under room temperature, the strength enhancement with loading rate is of minor interest. Therefore, the specimens in Group HWC were tested at a quasi-static strain rate ($10^{-5}sec^{-1}$).

STRENGTH CHARACTERISTICS

As discussed in Section 2, tension tests and compression tests were conducted on dog-bone and cube specimens, respectively. The tensile strength(f_t) wasdetermined from the maximum axial stressmeasured during each test. The compressive strength (f_c) and the corresponding splitting tensile strength (f_s) were determined to be the maximum axial stress and the stress corresponding to the maximum strain prior to a sudden drop on the load-displacement curve.All test results arepresented in Table 4.

Table 4. Tensile and compressive strengths of concrete

Group	Strain rate (sec^{-1})	Number of specimens	f_t (MPa)	f_c (MPa)	f_s (MPa)
HDR	10^{-5}	5	2.21	32.8	3.21
	10^{-4}	3	2.39		
	10^{-3}	3	2.79		
	10^{-2}	6	2.87		
	10^{-1}	4	3.40		
	$10^{-0.3}$	3	3.93		
LDR	10^{-5}	4	1.18	17.9	1.94
	10^{-4}	3	1.36		
	10^{-3}	3	1.44		
	10^{-2}	3	1.54		
	10^{-1}	5	1.82		
	$10^{-0.3}$	4	1.88		
HWR	10^{-5}	3	1.30	24.46	2.45
	10^{-4}	3	1.59		
	10^{-3}	3	1.82		
	10^{-2}	3	2.20		
	10^{-1}	3	2.70		
	$10^{-0.3}$	3	3.07		
HDC	10^{-5}	3	2.53	33.3	3.76
	10^{-3}	4	2.93		
	10^{-1}	4	3.79		
HWC	10^{-5}	4	6.32	40.2	6.47

According to Bischoff and Perry (1991) and Yan and Lin (2006), the ***DIF*** may be expressed as a linear function of strain rate on a semi-log scale. That is,

$$DIF = 1 + \alpha \log(\dot{\varepsilon}_t / \dot{\varepsilon}_{ts}) \tag{1}$$

in which ***DIF***= f_t/f_{ts}, f_t denotesdynamic tensile strength at $\dot{\varepsilon}_t$, f_{ts}represents the quasi-static tensile strength at $\dot{\varepsilon}_{ts}$, α is a regression coefficient, $\dot{\varepsilon}_t$ is the strain rate above 10^{-5} sec^{-1}, and $\dot{\varepsilon}_{ts}$ is the quasi-static strain rate or $10^{-5}sec^{-1}$ in this study. Other values ranging from $10^{-5}sec^{-1}$ to $10^{-8}sec^{-1}$ have been used in the literature to represent the quasi-static strain rate.For example, Malvar and Ross (1998) assumed $\dot{\varepsilon}_{ts}$=$10^{-6}sec^{-1}$. Comite Euro-international du Beton (CEB) model code curves (1993) start at 3×10^{-5} sec^{-1}, which is close to the ASTM C496 recommendation for a standard tensile splitting test. In the region of $10^{-6}sec^{-1}$ to $10^{-5}sec^{-1}$, the ***DIF*** is nearly independent of strain rate.

By least square linear fitting of the experimental data, the regression coefficient α can be determined for Group HDR, LDR, HWR, and HDC as given in Table 5.

Table 5. Parameter in Eq. (1)

Group	HDR	LDR	HWR	HDC
α	0.134	0.135	0.265	0.115

Note that the general trend of strengthincrease with strain rate is similar to that given by Malvar and Ross (1998), which was derived from the test results of relatively small specimens. Due to presence of the free water in concrete, the viscous resistance on the strain rate sensitivityappears significant. For example, a simple comparison on the αcoefficients for Group HDR and HWR indicates that the strength increase in fully saturated concrete is much greater than that experienced by concrete with normal moisture content. This phenomenon quite differs from that observed under quasi-static loads. The static strength decreases in fully saturated concrete 1.30MPa as compared to the concrete with the normal moisture content 2.21MPa, as shown in Table 4. This is due to the fact that the presence of moisture forces the gel particles apart and reduces the van der Waals forces (Ross et al., 1996).

By comparing Group HDR and LDR tests, Table 4 also indicates that the concrete compressive strength has a minimum influence on the strain rate sensitivity. These results differ from those by Cowell (1966), where concrete with the lower compressive strength exhibited a higher ***DIF*** in tension. Cowell (1966) explained that, less dense than the higher strength concrete, the lower strength concrete generally possessed more micropores and thus has more pronounced viscous resistances, resulting in a higher ***DIF*** in tension under high strain rate loading. However, this is not always the case. In this study, the mix designs in Table 1 and the aggregate size of the concrete have been optimized to achieve the predetermined high density; the lower strength concrete is nearly as dense as that of the higher strength concrete. As a result, the lower strength concrete has nearly the same ***DIF*** as that of the higher strength concrete.

Table 6compares the current study on Group HDR specimens with Malvar and Ross (1998) and CEB (1993). In general, the CEB model code gives a conservative estimate on the dynamic strengthat high strainrates. This study presents the highest dynamic strength associated with strain rates.

Table 6. Formulas on the dynamic strength enhancement

Strain rate (sec^{-1})	10^{-5}	10^{-4}	10^{-3}	10^{-2}	10^{-1}	1
CEB model code (1993)	1.000	1.082	1.171	1.267	1.371	1.483
Malvar and Ross (1998)	1.000	1.088	1.184	1.289	1.402	1.526
Eq. (1) for Group HDR	1.000	1.134	1.268	1.402	1.536	1.670

Three representative modes of tension failures are shown in Figure 13 under various strain rates.

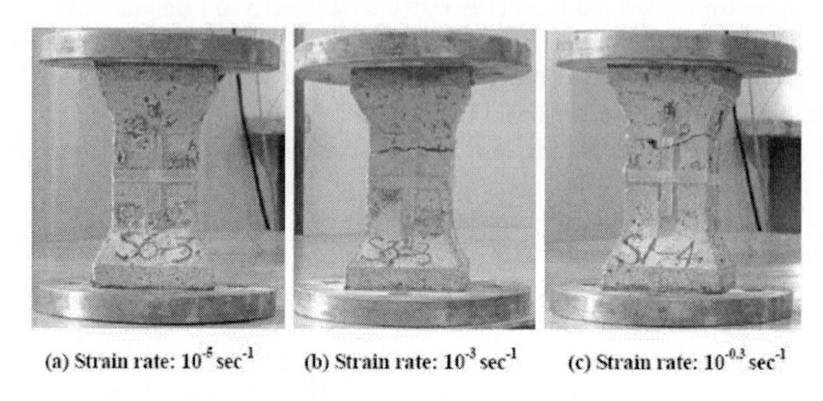

(a) Strain rate: 10^{-5} sec^{-1} (b) Strain rate: 10^{-3} sec^{-1} (c) Strain rate: $10^{-0.3}$ sec^{-1}

Figure 13. Typical failure modes of concrete specimens.

DEFORMATION AND ENERGY ABSORPTION CAPACITY

The occurrence of cracks on the surface of a test specimen gradually makes the deformation of the specimen become non-uniformly distributed. The reading from a strain gauge or an LVDT represents an average strain or displacement within its measurement coverage and is thus more meaningful prior to the onset of cracks.Representative stress-strain curves of the test specimens are presented in Figure 14for the saturated concrete specimens in Group HWR. It can be clearly observed from the figure that not only the peak stress of each curve but also the strain corresponding to the peak stress and the slope of each curve increase with strain rate. These trends will be quantified in the following sections.

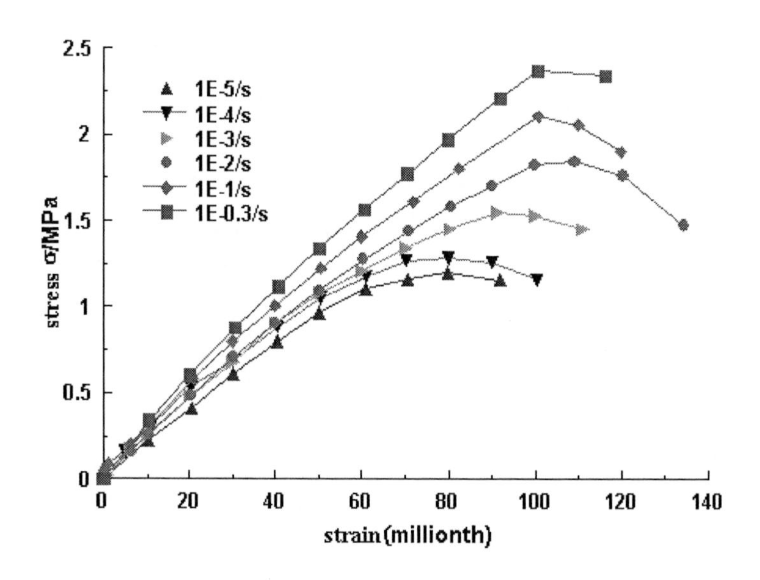

Figure 14. Typical stress-strain curves of saturated concrete (Group LWR).

Modulus of Elasticity

The modulus of elasticity is typically calculated as either a secant or a chord modulus. A secant modulus is calculated from the origin to a defined point on the stress-strain curve, usually within 30 to 60% of the sample's ultimate strength. The chord modulus is usually evaluated according to ASTM C 469 between the stress and strain pairs at 50 millionths strain and at 40% of the ultimate strength. In this study, the dynamic and static elastic modules were measured as the chord modulus according to ASTM C 469-02.

Many researchers have reported that a high modulus is obtained under rapid loading (Ross et al., 1995; Reinhardt et al., 1990; Rossi and Toutlemonde, 1996). At high load rates, the viscous resistance of free water in concretemicroporesand the inertial resistance of coarse aggregates in concretedelay both the onset of microcracks and the propagation of initial microcracks. As a result, the modulus of elasticity increases. As pointed out by Rossi and Toutlemonde (1996), the viscous resistance, or the Stéfan effect has a greater influence on the strength than on the modulus of elasticity. Therefore, strain rate affects the modulus of elasticity to a much lesser extent than the tensile strength of concrete. In this study, the average moduli of elasticity from the quasi-static test of Group HDR, LDR and HWR specimens are 2.86×10^4

MPa, 2.03×10^4 MPa and 1.89×10^4 MPa, respectively. The relative increases in modulus of elasticity with strain rate are summarized in Table 7 in reference to a quasi-static strain rate of 10^{-5}sec^{-1}.

Table 7. Relative increase in modulusof elasticity

Group	Strain rate (sec^{-1})				
	10^{-4}	10^{-3}	10^{-2}	10^{-1}	$10^{-0.3}$
HDR	4.0%	9.0%	10.1%	11.4%	12.1%
LDR	1.3%	4.9%	9.3%	12.0%	18.0%
HWR	28.3%	30.8%	45.8%	70.7%	100.1%

Similar to Eq. (1) for strength increment, Eq. (2) is proposed to quantify the dynamic effect on the modulus of elasticity:

$$E_t / E_{ts} = 1.0 + \beta \log(\dot{\varepsilon}_t / \dot{\varepsilon}_{ts}) \quad (2)$$

Where E_t is the modulus of elasticity at high strain rates; and E_{ts} is the quasi-static modulus of elasticity at a strain rate of 10^{-5}sec^{-1}. By least square linear fitting, the regression coefficient $\boldsymbol{\beta}$ can be obtained as summarized in Table 8 for various groups of specimens.

Table 8. Parameterin Eq.(2)

Group	HDR	LDR	HWR
β	0.023	0.037	0.188

As described previously, the static modulus of saturated concrete (Group HWR) is lower than that of concrete with normal water content (Group HDR). Thisresult is similar to the relation for tensile strength as shown in Table 4. The presence of moisture forces the gel particles apart so that the static elastic modulus of saturated concrete (Group HWR) is lower than that of concrete with normal water content (Group HDR). Due to the viscous resistance of free water in micropores, the strain-rate effect on the modulus of elasticity of the saturated concrete becomes more significant than that of the normal concrete.

In comparison with Table 5, Table 8 indicates a less pronounced effect of strain rate on the modulus of elasticity than on the tensile strength. Although it affects the strength***DIF***as shown in Table 5, temperature has a negligible influence on the strain-rate dependence of E_t / E_{ts} based on the comparison between specimens in Group HDR and in Group HDC.As the effect of temperature on the strain-rate sensitivity of the modulus of elasticity is less reported in the literature, further studies are needed to improve the current understanding on this topic.

Critical Strain

Many investigators have studied the effect of strain rate on the critical strain that corresponds to the peak stress. Most of the studies, however,are limited to the compressive behavior of concrete. The experimental results obtained by different authors are contradictory. Some indicate an increase of the critical strain at peak stress and the others show a decrease with strain rate. Bazant and Oh (1982) suggested a slight increase of the critical strain with strain rate. In their summary of previous research works, Bischoff and Perry (1991) statedthat the change in critical compressive strain ranged from a decrease of 30% to an increase of 40% for strain rates from $10^{-5}sec^{-1}$ to $10sec^{-1}$. Most results showed that the average increase in critical strain varied from 10% to 30% for strain rates as high as $10sec^{-1}$after outliers with negative results have largely been removed.It was less than the average increase of 60% to 80% in the compressive strength and significantly less than the 250% increase expected for the critical tensile strain at comparable strain rates(Bischoff and Perry, 1991). The discrepancy on the strain rate dependence of critical strain may be due to variation in test methods, leading to different failure pattern, or simply because of deficiencies in the measurement technique. The SHPB test conducted by Grote et al. (2001) at very high strain rates ranging from $290sec^{-1}$ to $1500sec^{-1}$ showed that the critical compressive strain increased slightly with strain rate. The amount of increase may be within the range of experimental error.

In this study, the critical tensile strainwas found to increase withstrain rate for each group of the test results. As indicated in Figure 14, the higher the strain rate, the sharper the peak stress area of stress-strain curves. At the quasi-static strain rate, the critical tensile strains for Group HDR, LDR, and HWR specimens are 73.8×10^{-6}, 70.1×10^{-6} and 75.0×10^{-6}, respectively. The critical strain gradually increases with strain rate. By the least square linear regression, theirrelation can be expressed into:

$$\varepsilon_p / \varepsilon_{ps} = 1.0 + \gamma \log(\dot{\varepsilon}_t / \dot{\varepsilon}_{ts}) \quad (3)$$

Wherein ε_p denotes the strain corresponding to the peak stress at strain rate $\dot{\varepsilon}_t$; ε_{ps} denotes the strain corresponding to the peak stress at quasi-static strain rate $\dot{\varepsilon}_{ts}$ or $10^{-5}sec^{-1}$ in this study. The regression coefficient in Eq. (3) is given in Table 9 for various groups of the test specimens.

Table 9. Parameterin Eq. (3)

Group	HDR	LDR	HWR
γ	0.203	0.109	0.067

A comparison between parameters in Table 8 and Table 9 shows an opposite tendencyof the strain-rate effecton the critical strain from that on the modulus of elasticity in two different temperature environments. This may be due in part to the fact that the critical strain is mainly affected by the static strength (Bischoff and Perry, 1991), and the presence of moisture reduces the static strength (Ross et al., 1995).

Comparing the parameterγ for Group HDR and LDR indicates that high strengthconcretehas a greater critical strain than low strength concrete.This observation agrees with the point of view of Bischoff and Perry (1991) on the compressive behavior of concrete: a stronger material should exhibit greater deformability.

For the same reason as for the modulus of elasticity, temperature exerts an insignificant influence on the strain-rate sensitivity of critical strain. Therefore, the γparameter is not given separately for Group HDC specimens. This may also have something to do with the fact that the static strength of Group HDC (33.3MPa)is close to that of Group HDR(32.8MPa)according to Table 4. In addition, further studies along this line may be needed since the factors that affect strain-rate dependence of critical strain are not well known.

Poisson's Ratio

Each specimen from Group HDR, LDR, HWR and HDC was instrumented with both longitudinal and transverse strain gages in order to assess deformability in both directions. Based on a complete set of the test datacollected from all groups of specimens, the dependence of Poisson's ratio upon strain rate is unclear. For example, the Poisson's ratios obtained from Group HDRspecimens vary from 0.12 to 0.18 and they seem to be independent of strain rate. Thus anaverage Poisson's ratio of 0.16 is suggested for Group HDR. Similarly, the Poisson's ratios for Group LDR, HWR and HDC are suggested as listed in Table 10.

Table 10. Suggested Poisson's ratio

Group	HDR	LDR	HWR	HDC
Poisson's ratio	0.16	0.15	0.17	0.15

Note that little data and information concerning the strain-rate effect on Poisson's ratio are available in the literature. CEB recommendations (1990) assumed that the Poisson' ratio is independent of loading rate simply due to lack of test results. Takeda and Tachikawa (1962) noticed an increase ofthe Poisson's ratio of concretewhen tensionedunder rapid loading. Paulmann and Steinert (1982) did not observe any change in Poisson's ratio for strain rate up to 0.2sec^{-1}. Based on the test data in this study, the effect of strain rate on the Poisson's ratio can be neglected in a strain rate range from 10^{-6} to 10^{-1} sec^{-1}.

Energy Absorption Capacity

Energy absorption capacity can be defined as the area under the stress-strain curve up to the peak stress level. In comparison with the results of quasi-static tests, there is a considerable increase in energy absorption capacity at high stain rates. The relative increase at high strain rates are summarized in Table 11. The increasing tendency is attributed primarily to the fact that due to the enhanced behavior of concrete at high strain rates,concrete fracture is forced to propagate through the region of high resistance. For example, some cracks that typically occur along aggregate-mortar interfaces under static loading may now cut through aggregates under dynamic loading, thus requiring a higher stresslevel with its corresponding similar strain level as discussed in Section 3.2.2to fracture a test specimen, or increasing the energy absorption capacity of the specimen.

Table 11. Relative increase in energy absorption capacity

Group	Strain rate (sec^{-1})				
	10^{-4}	10^{-3}	10^{-2}	10^{-1}	$10^{-0.3}$
HDR	1%	55%	63%	70%	91%
LDR	30%	31%	47%	68%	74%
HWR	28%	54%	78%	70%	134%

Discussion on Dynamic Failure Mechanism

Significant efforts have been dedicated to the research on the failure mechanism of concrete under dynamic loading. To account for the strain-rate effect on concrete behavior, various assumptions have been proposed by several investigators. Rossi and Toutlemonde(1996) concluded that, up to a strain rate of $1sec^{-1}$, dominant to the failure of concrete is viscous mechanism similar to the Stefan effect. At a strain rate of over $10sec^{-1}$, the forces of inertia become preponderant. In between, both viscous and inertial mechanisms may be significant. While Sercombe and Ulm et al. (1998) emphasized the viscous effect in dynamic behavior of concrete, Harsh et al. (1990) and Yan and Lin (2006) investigated the dynamic fracture behavior of concrete in compression and tension, respectively. Hence, more research worksare required before general conclusions can be drawn about failure mechanism under dynamic loading. In addition, the effect of initial static stress on the dynamic propertiesof concrete is yet to be further characterized. To better understand the mechanical behavior of the material, more studies must be conducted to investigate a broader range of strain rates and the effect of load paths encountered in various applications of interest.

Comborde and Mariotti (1999) simulated the dynamic failure mechanism of concrete with the discrete element method. They found that without the viscous effect, the inertia effect cannot account for the general behavior of concrete under dynamic loading. The confining pressureresulted from inertia effects could improve the compressive strength of concrete. The effect of inertia on the tensile strength of concrete is still unclear, as illustrated in Figure 15.

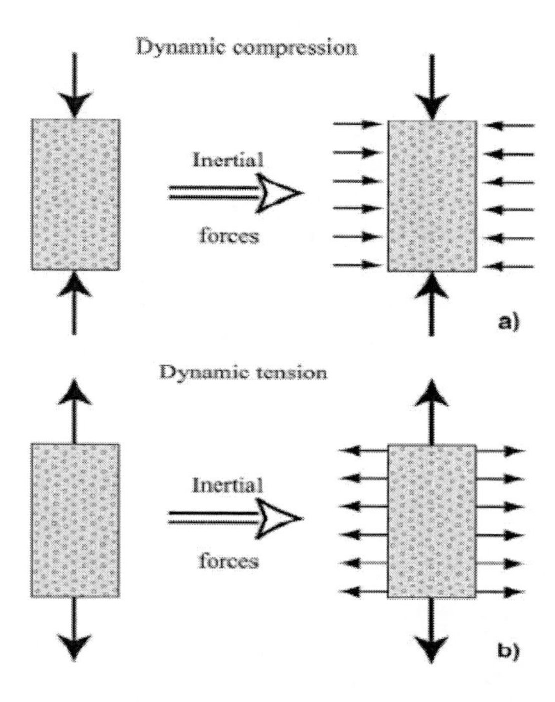

Figure 15. Effect of inertial confinementon uniaxial compressiveand tensile strengths.

Malvar and Ross (1998) reviewed the strain-rate effect onthe tension properties of concrete. They concluded that ***DIF*** may be expressed as a bilinear function of strain rate in a log-log plot with a slope change at a strain rate of $\dot{\varepsilon}_c = 1.0 \sec^{-1}$. Above the critical point, a more rapid increase in tensile strength was observed. This conclusion somewhat differs from the CEB model code (1993), where $\dot{\varepsilon}_c = 30 \ \sec^{-1}$. In reality, the change in slope may not occur at a definite point, but rather in a transitional region that depends upon many factors. Zielinski et al. (1981) tested both wet and dry concrete and found that there was no definite change of the dynamic tensile strength when the moisture condition changed. Other investigators reported that the increase in strength is greater in wet concrete than in dry concrete. Ross et al. (1996) attributed this moisture effect to the presence of free water contained in micropores of concrete, which caused the increase of strength at high strain rates. Several hypotheses concerning the physical mechanisms of strain-rate enhancement have been proposed. Founded upon experimental and analytical works on microconcrete, Rossi and Toutlemonde (1996) concluded that the viscous effects together with the forces of inertia also led to an increase of the modulus of elasticity of concrete but to a much lesser extent than the tensile strength. Cadoni et al. (2001) tested large-size concrete cubes of 20cmeach side with aggregates from powder to 25mmin maximum dimension. The specimens were subjected to various curing conditions to study the effect of internal humidity conditions on the strain-rate enhancement. They confirmed that the level of free water inside the concrete played an important role in the sensitivity of concrete response. In addition, they proposed a wave propagation concept to explain the physical phenomenon. They found that a stress wave can be considerablyamplified when it runs across dry voids and microcracks in concrete, but affected little by the voids filled up with free water.

In this study, laboratory testsindicated that the fractured surfaces of failed specimens became flatter as the strain rate increases and they cut through an increasing number of coarse aggregates. Representative modes of failureare depicted inFigure 16.

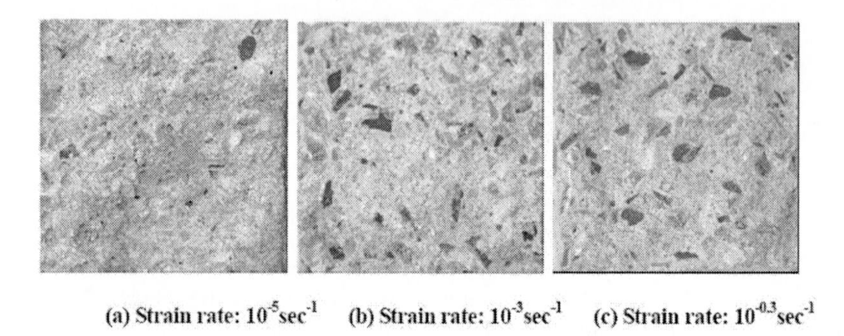

Figure 16. Typical fracture surfaces at different strain rates.

The experimentally observed strain-rate enhancements can be physically explained based on the microstructure and properties of concrete. Concrete is a composite material mainly consisting of sands and aggregatesof various sizes that are embedded in a cement paste matrix. Current research shows that a large number of bond microcracks exist at the interfaces between coarse aggregates and mortar (Hsu et al., 1963; Milashi et al., 1979). Uponloading, some of the microcracks can be developed due to the difference in stiffness between aggregates and mortar. Thus, the aggregate-mortar interface constitutes the weakest link in the composite system. In case of quasi-static loading, the failure is closely associated with the mechanism of internal progressive microcracking. At first, the bond cracks at the aggregate-mortar interface start to extend owing to stress concentration at the tip of the cracks.

Then, some cracks at nearby aggregate surfaces start to bridge in the form of mortar cracks, while other bond cracks continue to grow slowly. Finally, the microcracks through the mortar join together, resulting in a complete disruption of the concrete. The fracture surface of the specimen mainly passes through the mortar and the aggregate-mortar interfaces, leaving behind a rough surface as seen in Figure 17(a). However, the crackvelocity in concrete has been shown experimentally to increase with strain rate (Ross et al., 1996). Therefore, at low strain rates, crack has sufficient time to seek the path of least resistance. However, under rapid loading, the creation of new cracks is forced to propagate through the regions of high resistance, and a greater number of microcracks may be required before a continuous fracture surface can be formed. Thus, a certain part of coarse aggregates are broken during loading and the fractured surfaces of the specimens become flatter as seen in Figure 16(b) and (c), and Figure 17(b). The higher the loading rate is, the more fractured coarse aggregates are.

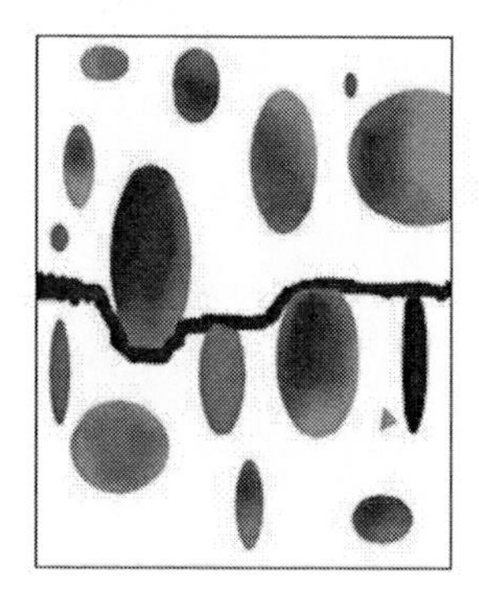

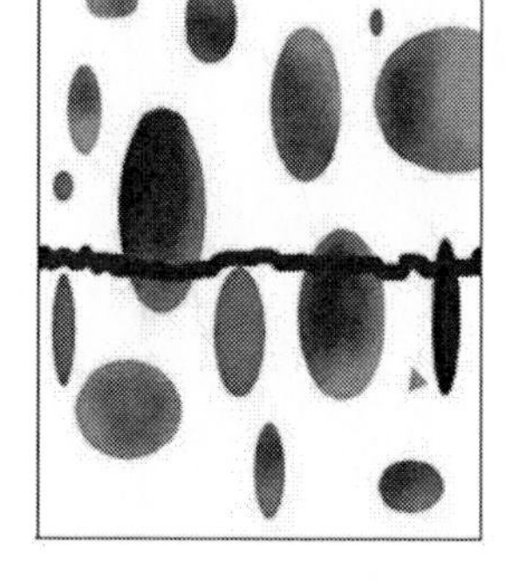

(a) Low strain rate (b) High strain rate

Figure 17. Illustration of failure mechanism of concrete in tension.

SUMMARY ON CONCRETE BEHAVIORS UNDER UNIAXIAL DYNAMIC LOADING

Based on the extensive tests and data analyses, the main findings in this section can be summarized as follows:

- The overall trend of strengthincrease with strain rate is similar to that given in the literature. However, the extent of strength increase observed from this study is more significant. The presence of free water is mainly responsible for more strength increaseinsaturated concretethan in dry concrete. At low temperature of -30°C, the increase in concrete strength is less sensitive to strain rate than in a room temperatureenvironment.

- The strain-rate enhancement for the modulus of elasticity is less pronounced than that for tensile strength. Temperature has little effects on the strain-rate sensitivity of both the modulus of elasticity and the critical strain. The moisture effects on the strain-rate sensitivity of the critical strain and the elastic modulus are opposite.
- Higher strengthconcretehas a larger critical strain than low strength concrete. No definite change of the Poisson's ratio was observed as the strain rate increases. The energy absorption capacity of concrete significantly increases with strain rate for both saturated concrete and concrete with normal water content.
- With the increase of strain rate, cracks tend to propagate through the regions of high resistance and cut aggregates into pieces. As a result, it takes a larger force to fracture a concretespecimen to failure under a higher strain rate.

Chapter 4

CONCRETE UNDER BIAXIAL DYNAMIC LOADING

For concrete structures in biaxial stress states, proportional biaxial loading and axial loading with constant lateral confinement are two representative load paths. They are of great significance to the understanding of concrete behaviors under arbitrary biaxial loading with the concrete unconstrained along the third axis (perpendicular to the other two axles). As stress increases along one axis, concrete tends to dilate along the other two axles. The custom-made test device in Figure 10 allows for simultaneously generating and monitoring the change of independent stresses along three axles. It can operate at a frequency of up to 10Hz, within which the lateral-to-axial stress ratiocan be kept at a specified value as the axial stress rapidly increases to study load-rate effects on concrete behavior.

STRENGTH CHARACTERISTICS

Table 12 presents the dynamic ultimate strengths of concreteunder compressive loading with various stress ratios. Each value in the table represents an average of at least four specimens. In general, the ultimate strength of concrete in biaxial compression state is higher than the uniaxial strength at any given strain rate due to the effect of lateral confinement. At a specified strain rate, the strength increment depends on the biaxial stress ratio. The maximum biaxial strength occurs at a stress ratio between 0.5 and 0.75 for any strain rate investigated in this study. In quasi-static tests, the ultimate strength of concrete tested under a stress ratio of (σ_1:σ_2 =1:0.75) is approximately 67% higher than the uniaxial strength, which is larger than the results obtained by Kupfer and Gerstle (1973) and Lee et al. (2004). The differences may be attributed to the slight difference in concrete mix design, boundary condition and loading path.

Table 12. Dynamic strength of concrete under biaxial stress states

Strain rate (sec^{-1})	Ultimate strengths at various stress ratios (σ_1:σ_2)				
	1:0	1:0.25	1:0.5	1:0.75	1:1
10^{-5}	9.84	14.86	16.13	16.39	14.00
10^{-4}	10.63	15.48	16.68	16.75	15.32
10^{-3}	11.38	16.17	17.36	17.54	16.66
10^{-2}	12.32	17.15	18.24	18.66	18.01

The ultimate strength at any stress ratio generally increases with strain rate. The strength increment at various stress combinations is quite different. In order to formulate the relationship between the strength increment and strain rate, several empirical relationships have been tried to fit the tested data. It was found that a linear logarithmic relation is rather simple and fits the test data well. That is,

$$\frac{f_{bd}}{f_{us}} = a + b\log(\dot{\varepsilon} / \varepsilon_s) \quad (4)$$

Where $\dot{\varepsilon}_s$ is the quasi-static strain rate, $10^{-5}sec^{-1}$; $\dot{\varepsilon}$ is the general strain rate; $\boldsymbol{f_{us}}$ is the uniaxial strength of concrete in quasi-static loading; $\boldsymbol{f_{bd}}$ is the dynamic strength of concrete in biaxial stress state; ***a***, ***b***are parameters depending on the material behavior and stress ratio.

The increase in uniaxial compressive strength at various strain rates is compared in Figure 18with those published by other researchers. It can be seen from Figure 18 that CEB model code (1993) gives relatively conservative values.

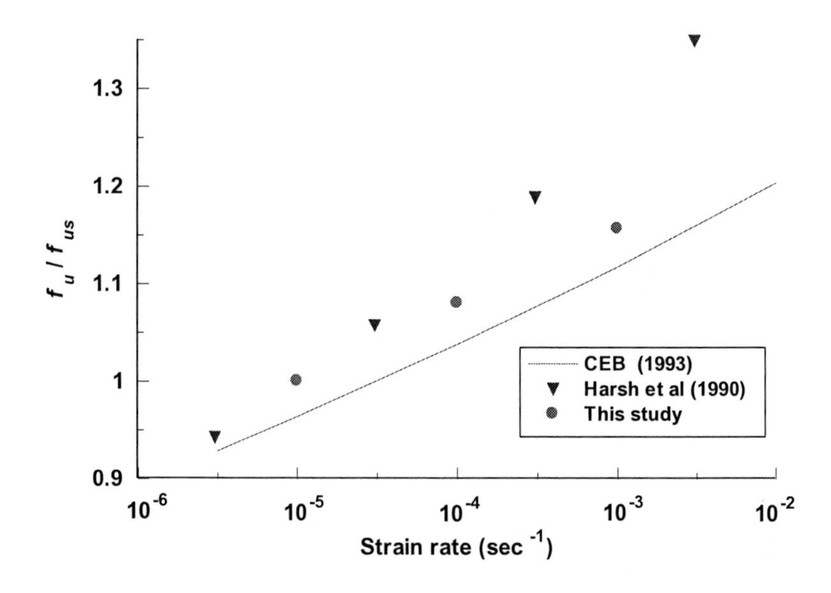

Figure 18. Uniaxial compressive strength with strain rate.

In this study, the coefficients ***a*** and ***b***were determined by the least square method fitting Eq. (4) to the test data. They are given in Table 13.

Table 13. Parameters at various stress ratios

$\sigma_1 : \sigma_2$	*a*	*b*	R^2	Mean error /MPa
1:0	1.00	0.0829	0.948	0.165
1:0.25	1.50	0.0769	0.834	0.308
1:0.5	1.63	0.0715	0.795	0.321
1:0.75	1.65	0.0767	0.774	0.609
1:1	1.42	0.136	0.853	0.507

If the compressive strength of a specimen under biaxial stress state is normalized by the uniaxial strength, the strain-rate effect is controlled by the lateral-to-axial stress ratio. Figure 19presents the normoalized strength versus strain rate relation for concrete under compression. It is observed from Figure 19 that the strain-rate sensitivity of concrete is more pronounced at stress ratios of 1:0 and 1:1.

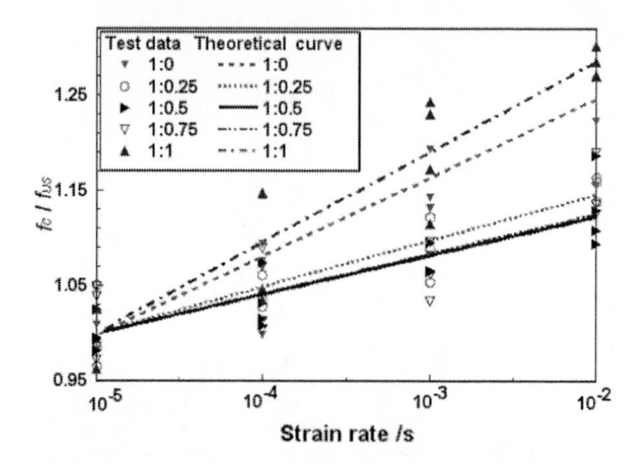

Figure 19. DIF at different stress combinations.

Kupfer and Gerstle (1973) proposed an empirical relation that represented the strength enhancement of concrete under biaxial stress states. The relation in a dimensionless form can be expressed into:

$$\frac{f_{bd}}{f_{ud}} = \frac{(c + d\alpha)}{(1+\alpha)^2} \tag{5}$$

where $\alpha(=\boldsymbol{\sigma_2/\sigma_1})$ is the stress ratio; $\boldsymbol{f_{ud}}$is the uniaxial strength at the current strain rate; $\boldsymbol{f_{bd}}$ is the dynamic strength of concrete in biaxial stress state; $\boldsymbol{c}$ and $\boldsymbol{d}$ are parameters determined by fitting the relation to the test data, as listed in Table 14.

Table 14. Parameters in Eq. (5)

Strain rate (sec^{-1})	c	d	R^2	Mean error (MPa)
10^{-5}	1.021	5.196	0.9172	0.531
10^{-4}	1.012	4.947	0.9484	0.358
10^{-3}	0.9994	4.897	0.9640	0.368
10^{-2}	0.9899	4.796	0.9588	0.358

Table 12 clearly indicated that the strength enhancement of concrete in biaxial stress states is a function of both strain rate and lateral confining pressure. A simple expression for the dynamic strength of concrete in biaxial stress statescan be formulated by combining Eqs. (4) and (5):

$$\frac{f_{bd}}{f_{us}} = P_1 + P_2 \log(\dot{\varepsilon} / \dot{\varepsilon}_s) + \frac{P_3}{(1+\alpha)^2} + \frac{P_4 \alpha}{(1+\alpha)^2} \quad (6)$$

wherein $\boldsymbol{P_1}$,$\boldsymbol{P_2}$, $\boldsymbol{P_3}$ and $\boldsymbol{P_4}$ represent the parameters associated with material properties. By fitting to the test data, $\boldsymbol{P_1}$,$\boldsymbol{P_2}$, $\boldsymbol{P_3}$,$\boldsymbol{P_4}$are determined to be -0.446, 0.0875, 1.43 and 6.42, respectively. The correlation coefficientof Eq. (6) with the test data in Table 12 is 0.958 and their mean error is 0.412MPa. To visualize the accuracy of Eq. (6), Figure 20 compares the predictions with the test results. Clearly, a good agreement between the two has been achieved.

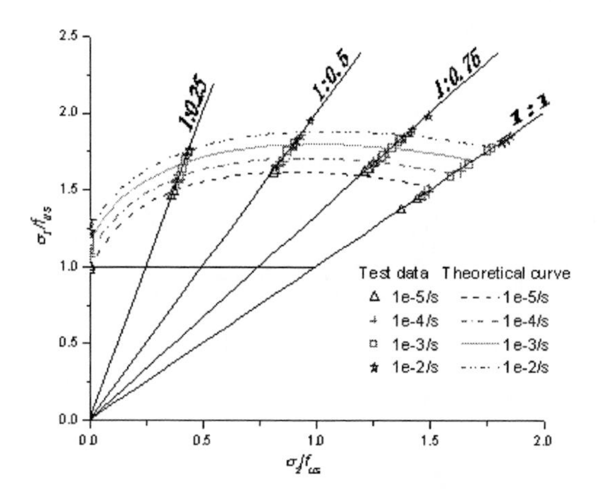

Figure 20. Biaxial strength envelop for concrete under biaxial compressive stresses.

It should be noted that the proposed strength relation, Eq. (6), takes into account the strain rate effect and the confining pressure effect independently, and thus it is unable to reasonably reflect the coupled effect of strain rate and lateral pressure. Considering the scattering data, the proposed strength relation is deemed acceptable.

DEFORMATION CHARACTERISTICS

The force and deformationalong the two loading axle were measured for each specimen. In this book, the strain sign convention is defined such that compressive strain is positive and tensile strain is negative.

Figure 21 shows typical stress-strain curves of concrete at various strain rates. It can be observed from Figure 21 that, at a specified stress ratio, the shapes of the stress-strain curves corresponding to various strain rates are similar. The effective modulusof elasticity(i.e. the slope of a stress-strain curve) for confined specimens is slightly higher than that for unconfined ones. Up to a stress level of approximately 60% of the peak stress, the stress-strain relation is basically linear. As the strain rate increases, the linear part providing the tangent modulus of elasticity extends. However, the increase in tangent modulus in the inelastic rangebecomes more pronounced with rising strain rate.

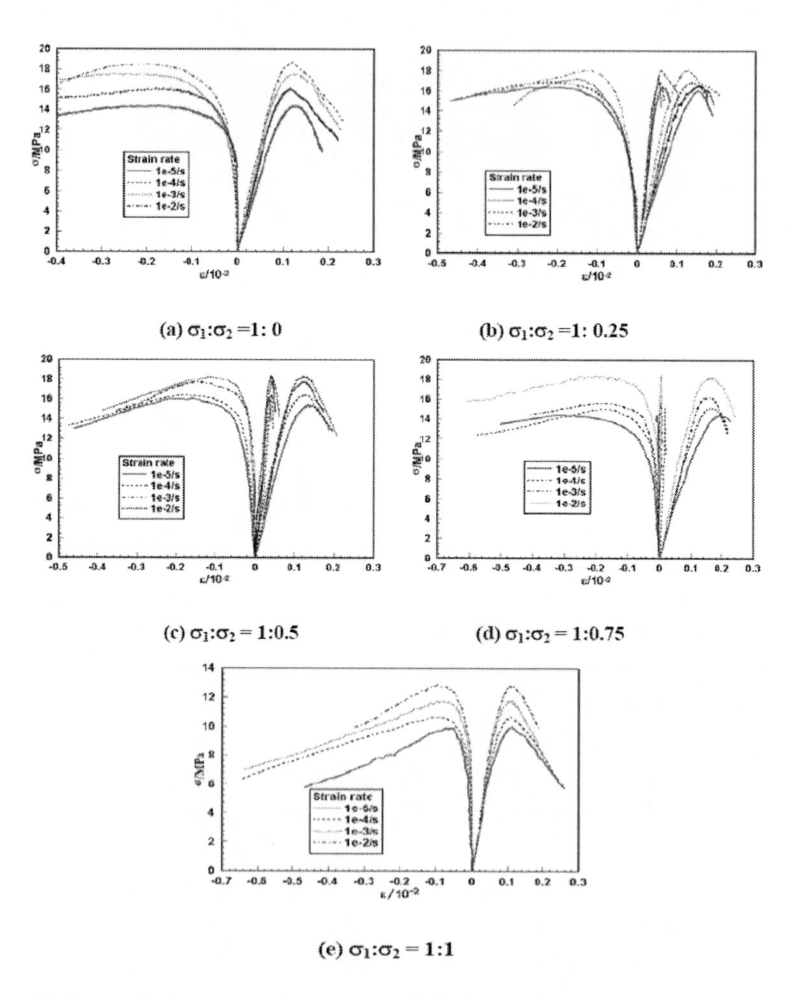

Figure 21. Stress-strain curves of concrete at different strain rates.

FRACTURE CHARACTERISTICS

The crack patterns and failure modes of tested specimens observed during various dynamic tests are very similar to those in static experiments (Kupfer and Gerstle, 1973; Lee and Song et al., 2004). There is no clear indication of their dependence on strain rate. Figure 22shows the failure modes of four specimens tested under a stress ratio of 1:0.5 at various strain rates. No fundamental differences were observed between them. However, with the increasing of strain rate, slightly more microcracks may develop aroundone or a few major cracks. In this case, slightly more coarse aggregateswere brokenbased on a close examination on the fracture sections. This might be attributed to the fact that at high strain rates, cracks may not have sufficient time to develop along the path of least resistance; they are forced to propagate through the regions of high resistance. As a result, a high peak stress was observed at the failure of specimens.

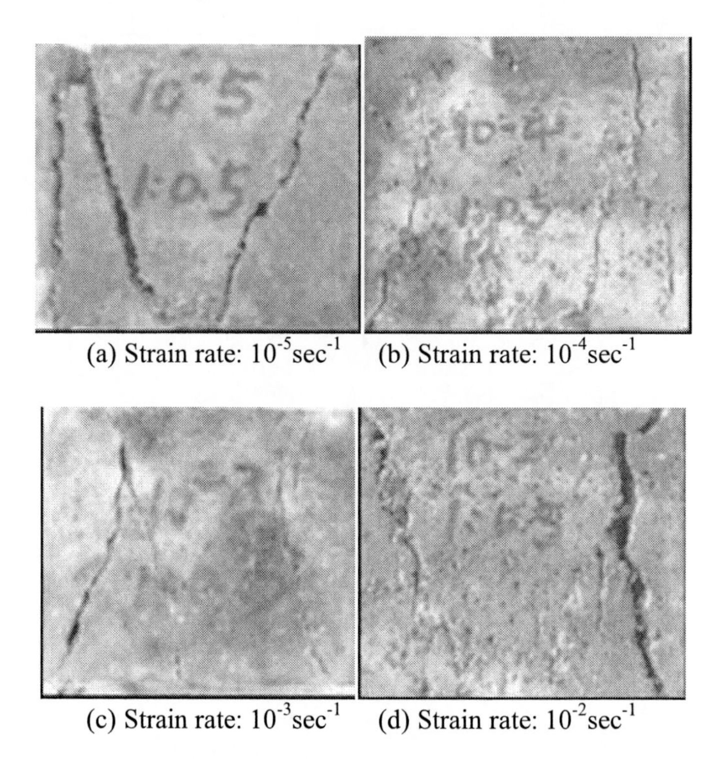

(a) Strain rate: $10^{-5}sec^{-1}$ (b) Strain rate: $10^{-4}sec^{-1}$

(c) Strain rate: $10^{-3}sec^{-1}$ (d) Strain rate: $10^{-2}sec^{-1}$

Figure 22. Typical failure modes of specimens at different strain rates (1:0.5).

Figure 22demonstrates the failure modes observed for various stress ratios at a strain rate of $10^{-2}sec^{-1}$. Under uniaxial compression, the fracture of the tested concrete specimens is characterized by the formation of a major crack and some minor cracks in parallel to the direction of the applied load. The fracture of the specimens tested under a stress ratio of 1:0.25, 1:0.5 and 1:0.75, respectively,is characterized by the formation of several major cracks that are in 10° ~45°angle to the direction of the applied load.

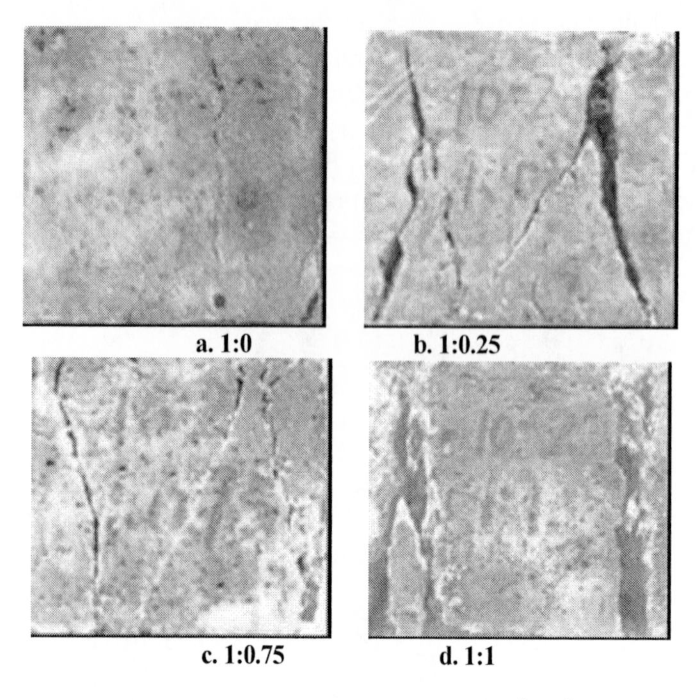

Figure 23. Fracture modes of specimens at a strain rate of $10^{-2}sec^{-1}$.

Summary on Concrete Behaviors under Biaxial Dynamic Loading

Based on the above test data and analysis, the following conclusions on the dynamic behavior of concrete in biaxial compressive states can be drawn:

- The dynamic strength of concrete confined in one lateral direction increases with strain rate. The extent of increment depends on the lateral-to-axial stress ratio.

- The proposed strength enhancement factor in Eq. (6) for concrete under biaxial stress statestakes into account the effects of both strain rate and lateral confinement. The established equation is in good agreement with the test data.
- The tangent modulus of elasticity of concrete or the slope of stress-strain curvesinitially starts to increase slightly with strain rate, but rapidly changes around the peak stress.
- The failure mode and crack pattern of specimens are predominantly controlled by the lateral-to-axial stress ratio and nearly independent of the strain rate.
- The strain-rate dependence of specimens in uniaxial tension and uniaxial compression has been shown similar when the specimens were cast with two concrete mix designs that give the 28-day compressive strengthof 10.7MPa and 21.2 MPa.

Chapter 5

CONCRETE UNDER TRIAXIAL DYNAMIC LOADING

STRENGTH CHARACTERISTICS

The confining pressure levels considered in this study were taken to be 0, 4, 8, 12 and 16 MPa. At each confinement level, the strain rate was considered to be 10^{-5} sec^{-1}, 10^{-4} sec^{-1} or 10^{-3} sec^{-1}. Table 15 gives the dynamic strengths of all test specimens. Each representsan average value of at least four specimens. Note that compressive stresses and compressive strains are defined positive in this study.

Table 15. Dynamic concrete strengths in triaxial stress states (MPa)

Confining pressure (MPa)	Strain rate (sec^{-1})		
	10^{-5}	10^{-4}	10^{-3}
0	9.84	10.63	11.38
4	30.05	32.11	33.70
8	46.27	48.08	49.39
12	61.21	61.42	61.16
16	72.14	75.34	74.08

Figure 24 compares the triaxial test data from this study and those by other investigators(Richart et al.,1928; Newman,1979; Imran and Pantazopoulou,1996;Sfer et al., 2002; Ansari and Li,1998; and Candappa et al., 1999). Under quasi-static loading, the triaxial strength of concretefrom all studies was found to significantly vary with the concrete confinement. According to Richart et al. (1928), the axial strength of test specimens is proportional to the confining pressure:

$$\frac{f_c}{f_{us}} = 1 + 4.1\frac{\sigma_{lat}}{f_{us}} \tag{7}$$

in which σ_{lat} is the lateral confining pressure and f_c is the ultimate strength in triaxial stress state. As shown in Figure 24, Eq. (7) can accurately predict the test data by Imran and Pantazopoulou (1996) and Sfer et al. (2002). However, in comparison with the test data acquired from this study, it underestimates the strength of the specimens at low confinement and slightly overestimates the strength at high confinement. This main difference is due primarily to the presence of lateral dilations and the use of different concrete materials in the previous studies. Newman (1979) also recommended a nonlinear relation between the ultimate strength and confining pressure. As shown in Figure 24, the nonlinear relation can represent the test results in this study better than the linear relation by Richart et al. (1928).

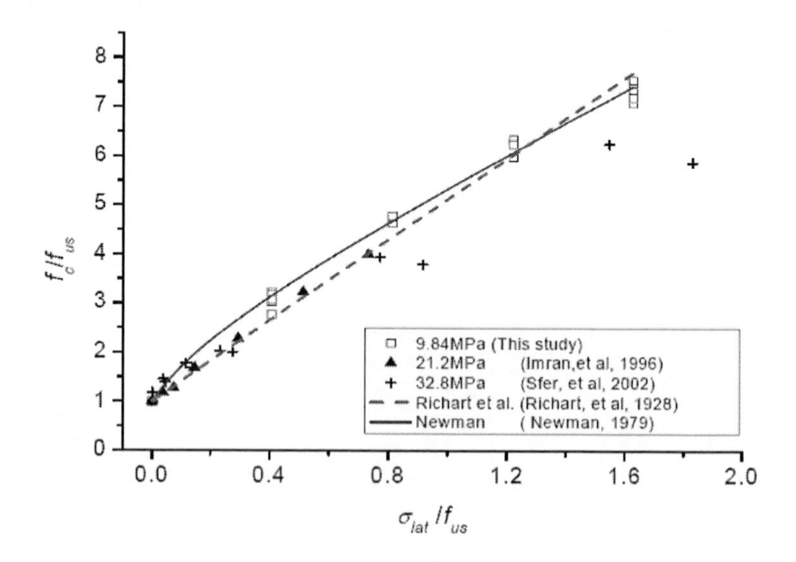

Figure 24. Strengths in quasi-static tests with existing failure envelopes.

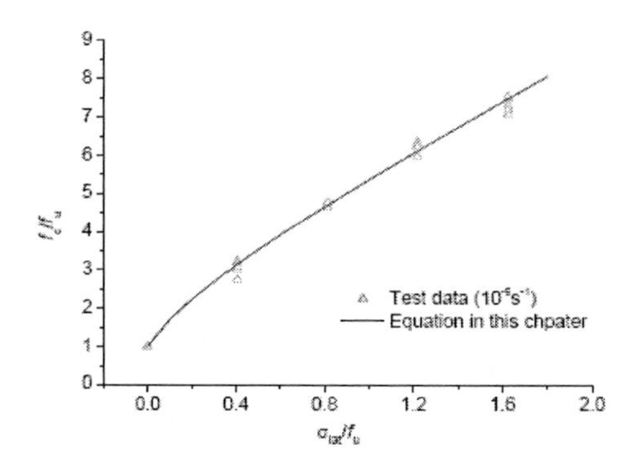

(a) Strain rate: 10^{-5} sec^{-1}

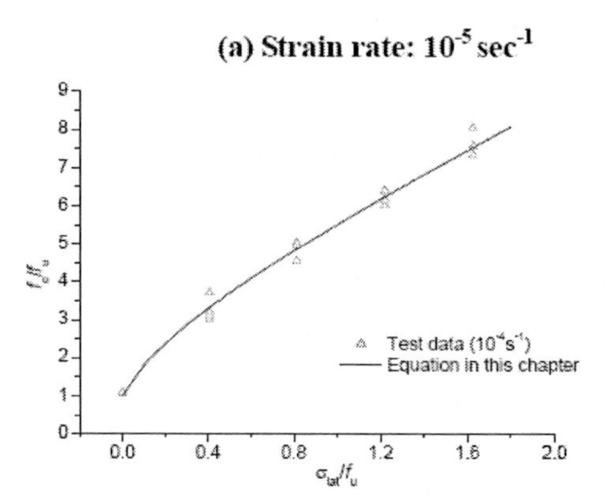

(b) Strain rate: 10^{-4} sec^{-1}

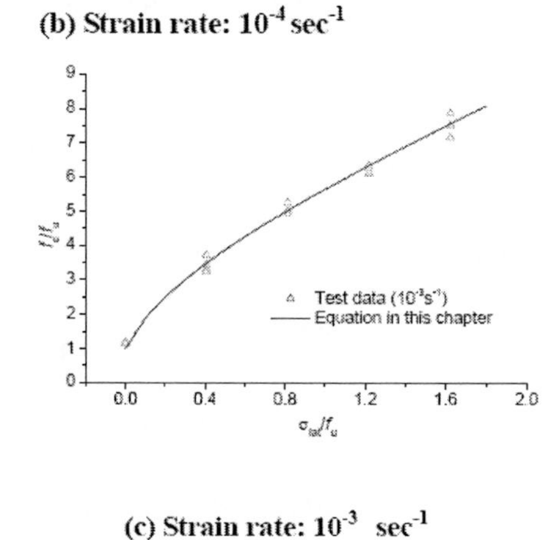

(c) Strain rate: 10^{-3} sec^{-1}

Figure 25. Triaxial strengths of concrete at various strain rates.

To take into account the effect of strain rates, the Newman's equation is modified into:

$$\frac{f_c}{f_u} = \sqrt{A(\dot{\varepsilon})\left(\frac{\sigma_{lat}}{f_u}\right)^2 + B(\dot{\varepsilon})\frac{\sigma_{lat}}{f_u} + 1} \tag{8}$$

where $A(\dot{\varepsilon}) = a + b\log(\dot{\varepsilon})$ and $B(\dot{\varepsilon}) = c + d\log(\dot{\varepsilon})$; $f_u$is the uniaxial compressive strength of concrete at strain rate $\dot{\varepsilon}$; the material constants ***a*,*b*, *c*** and ***d***were determined in this study to be 2.22, -1.54, 23.7 and 1.19, respectively. Figure 25 (a, b, c) shows the ultimate strengths at various strain rates as a function of the confining pressure. For comparison, Eq. (8) was plotted in Figure 25 as well. It can be seen from Figure 25 that Eq. (8) is in excellent agreement with the test data. It should be noted that the values predicted by the formula only make sense when confining pressure is smaller than the uniaxial static strength of concrete.

A comparison among Figure 25 (a), (b), and (c) indicates that the failure envelope in the principal stress space gradually expands with strain rate at low confining pressure (i.e.$\sigma_{lat}<f_{us}$). As the confining pressure further increases, the failure envelope appears insensitive to the strain rate of the applied loads.

DEFORMATION PROPERTIES

Stress-Strain Curves

Figure 26presents three sets of stress-strain curves in terms of axial strain, transverse strain and volumetric strain, based on various quasi-static tests at a low strain rate. In Figure 26(a), the positive strain represents the axial strain in compression while the negative strain means the transverse strain in tension. Each test was completed after concrete crushed or experienced little strength reduction due to hydrostatic pressure effects. It is noted that the axial stress includes the initial hydrostatic pressure and the axial strain does not exactly start from zero, which may result in some differences in comparison with the results of previous studies (Ansari and Li, 1998; Sfer et al., 2002).

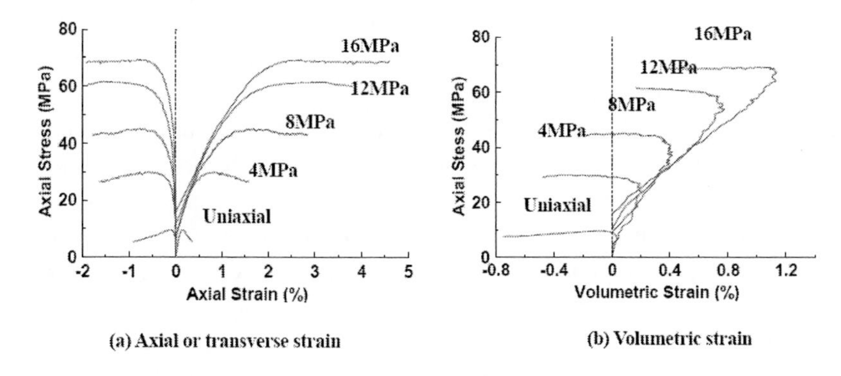

Figure 26. Stress-strain relationships at $10^{-5}sec^{-1}$.

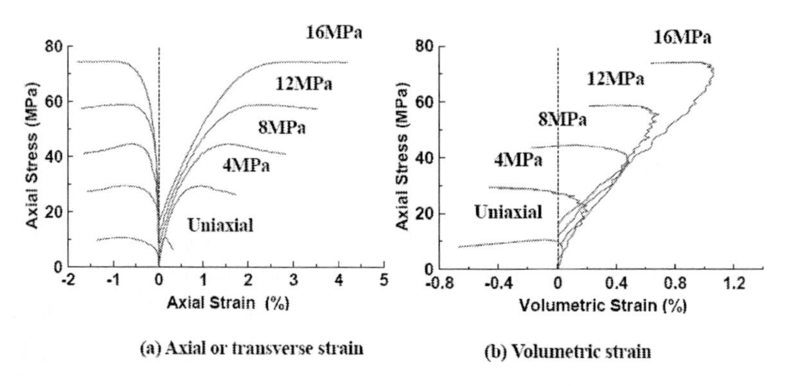

Figure 27. Stress-strain relationships at $10^{-4}sec^{-1}$.

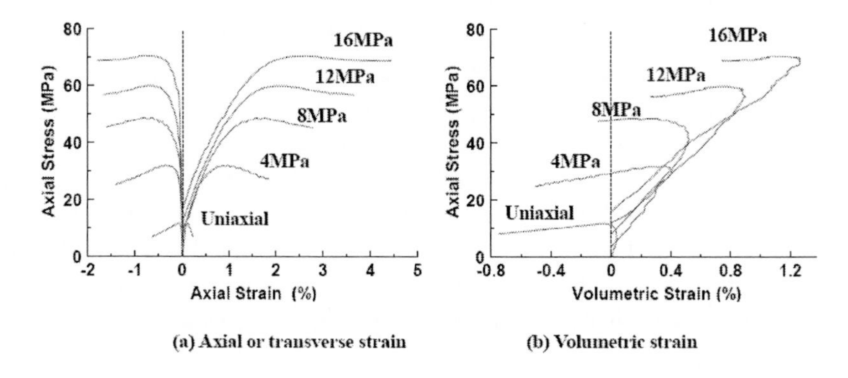

Figure 28. Stress-strain relationships at $10^{-3}sec^{-1}$.

As shown in Figure 26(a), the unconfined concrete specimen crushed at an ultimate strain of 0.4%. Confinement on concrete specimens can lead to significant differences in the shape of stress-strain curves. At low confining pressure (e.g. 4 MPa), the stress-strain curve exhibits a well-defined ultimate strength with a smooth descending, post-peak tail. As the confining pressure increases (up to 16 MPa), the stress-strain curves appear in a plateau at the concrete strength and exhibit little strength reduction over a large range of strains. This phenomenon likely results from the effect of hydrostatic pressures. Under a high hydrostatic pressure, concrete may have the microstructure that has been slightly damaged and thus experience a reduction in volume. This compression process does not cause a dramatic increase or decrease in strength and thus results in a plateau on the stress-strain curves. It is also observed that, as confinement increases, the initial slope of the stress-strain curve tends to decrease and the strain at the concrete strength increases dramatically.

Critical Strain

The axial stress is presented in Figure 26(a) as a function of the transverse strain at various confining pressures. Note that the transverse strain shown in Figure 26 (a) is actually an average of the two lateral strains. The results are very similar to those by Sfer et al. (2002) though the concrete used in this study had significantly lower strength.

The two lateral and one axial strains can be added together to obtain the volumetric strain, as shown in Figure 26(b), and expressed as a function of axial stress. It can be seen from Figure 26(b) that the material was initially compacted (i.e. volumetric reduction). The trend was then reversed immediately before the concrete strength has been achieved, exhibiting dilatancy (i.e. volumetric increase). In most cases, the ultimate strength of concrete was reached at approximately zero volumetric strain.

At high strain rates, the stress-strain curves of concrete specimens are presented in Figure 27 and Figure 28. In comparison with Figure 26, their shapes are similar, though the concrete strength is slightly increased. The strain corresponding to the concrete strength or maximum stress at different strain rates can be evaluated by the following linear regression equation:

$$\frac{\varepsilon_{1p}}{\varepsilon_{us}} = 1 + 18.6(\frac{\sigma_{lat}}{f_{us}}) \quad (9)$$

in which ε_{1p} and ε_{us} are the axial strains corresponding to the maximum stresses in triaxial stress state and uniaxial stress state, respectively. This equation was previously proposed by Ansari and Li (1998) and modified by Candappa et al. (2001), and then in this study. The correlation coefficient is 0.971, indicating that Eq.(9) can be used to accurately predict the measured strain.

SUMMARY ON CONCRETE BEHAVIORS UNDER TRIAXIAL DYNAMIC LOADING

The dynamic behavior of concrete in triaxial stress state has been studied under a linearly increasing load of high strain rates. Based on the extensive test data and analyses, the following conclusions can be drawn:

- The ultimate strength of concrete nonlinearly increases with its confining pressure at all load/strain rates that were considered in this study. As the confining pressure increases, concrete can endure significantly more deformation prior to crushing due to hydrostatic pressure effects.
- At low confining pressure, the ultimate strength of concrete increases with strain rate. When the confining pressure is close to or greater than the uniaxial static strength, the ultimate strength tends to slightly decrease with strain rate. The axial strain corresponding to the ultimate strength appears independent of strain rates.
- At various strain rates, the stress-strain curves of concrete are similar in shape. Their initial ascending portions have decreasing slopes as the confining pressure increases, implying the softening behavior of concrete.
- In triaxial stress state, concrete under compressionis first compressed and then expanded after the ultimate strength of concrete has been achieved.

Chapter 6

CONCLUDING REMARKS AND RECOMMENDATIONS

CONCLUDING REMARKS

In general, the ultimate strength of concreteincreases with strain rate. In both biaxial and triaxial stress states, the extent of strength enhancementtends to decrease with lateral confining pressure, indicating a reduced Stefan effect on the material strain-rate sensitivity. In the case of triaxial loading, the change in concrete strength becomes negligible as the confining pressure approaches or exceeds the uniaxial quasi-static strength of concrete. This is because the micro cracks gradually develops inside the concrete as the confining pressure increases, but rapidly expands when the confining pressure approaches the uniaxial quasi-static strength.

In a uniaxial state of tension or compression, concrete fails due mainly to propagation of fracture inside the paste, interfacial transition zone or aggregate. Under both biaxial and triaxial loading, the hydrostatic pressure effect may play an important role in the increase of concrete capacity.

As the confining pressure increases, concrete(particularly higher strength concrete) can endure significantly more deformation prior to crushing.In a triaxial stress state, concrete under compressionis first compressed and then expanded as the ultimate strength of concrete is gradually being achieved.The critical axial strain corresponding to the ultimate strength of concrete increases slightly with strain rate.

At various strain rates, the stress-strain curves of plain concrete are similar in shape. Their initial slope or the modulus of elasticity of concrete increases slightly with strain rate, but decreases with confining pressure. The Poisson's ratio of concrete changes little, the modulus of elasticity changes less than the concrete strength does as the strain rate increases. The area underneath a stress-strain curve or the energy absorption capacity of concrete significantly increases with strain rate.

Temperature has little effect on the strain-rate sensitivity of both modulus of elasticityand the critical strain corresponding to the concrete strength. Temperature slightly increases the strain-rate sensitivity of concrete strength. While it has a significantly less influence on the strain-rate sensitivity of the critical strain that varies with the static strength of concrete, moisture significantly increases that of the concrete strength and more so for the modulus of elasticity due to viscous resistance of the free water in concrete micropores.

Under rapid loading, cracks are forced to propagate through the regions of high resistance, thus occurring at a high stress level associated with significant energy absorption. As soon as cutting through coarse aggregates, cracks lead to concrete fracture. The higher the load/strain rate is, the flatter the fractured surfaces become. The failure mode and crack pattern of concrete specimensunder multiaxial loading are mainly controlled by the lateral-to-axial stress ratio and nearly not affected by strain rate.

Future research directions are directed to the in-depth understanding of concrete behaviors, such as the precise description of deformation under dynamic loading and the relevant mechanisms controlling various behaviors at high strain rates. In doing so, accurate constitutive relationships can be established and implemented into finiteelement analysis software so that a more reliable prediction of the behavior of concrete structures under dynamic loading conditions can be achieved.

RECOMMENDATIONS FOR CIVIL ENGINEERING DESIGN

It is recommended that different *DIF* values be used for concrete strength and the modulus of elasticity under uniaxial dynamic tension, and biaxial or trixial dynamic compression, respectively. The change in Poisson's ratio can be neglected.

It is recommended that the increasing effect with moisture on *DIF* values be considered. The temperature effect on the *DIF* is negligible.

It is recommended that the loading history be taken into account in the determination of *DIF* values for concrete structures under multiaxial loading. When the confining pressure is larger than 50% of the static strength of concrete, it is recommended that a reduced *DIF* be considered in practical design. For triaxial stress states, when the confining pressure equals or exceeds the static strength of concrete, a *DIF* of 1.0 must be taken.

ACKNOWLEDGMENT

Financial support to complete this research was provided by the National Science Foundation of China under Grant No. 50809064 at Zhengzhou University, China, and Nos. 50139010 and 90510018 at DalianUniversity of Technology, China. Special thanks go to Center for Infrastructure Engineering Studies at Missouri University of Science and Technology, USAfor its support to allow the authors to continue with their international collaboration and summarize the results and findings reported in this book.

The participation of Drs. Jun-yu Liu and Ping Xu in this research is highly appreciated. Grateful application is expressed to my wife, MrsQun Li who has provided numerous support and assistance to the first author.

REFERENCES

Abrams D A. Effect of rate of application of load on the compressive strength of concrete (Part 2). *ASTM Journal,* V. 17, 1917, pp. 364-377.

Adam S. Dynamic strength criterion for concrete. *ASCE Journal of Engineering Mechanics,* V. 130, No.12, 2003, pp. 1428-1435.

Ahmad S H, Shah S P. Behavior of hoop confined concrete under high strain rates. *Journal of ACI,* V. 82, No.5, 1985, pp. 634-647.

Ansari F, Li Q. High-strength concrete subjected to triaxial compression. *ACI Materials Journal,* V. 95, No.6, 1998, pp. 747-755.

Baishya M.C., Cook R.L. and Kelly M.T. Testing of polymer injection material, *Concrete International,* 1997, 19(4): 48–51.

Bazant Z.P. and Oh H. Strain-rate effect in rapid triaxial loading of concrete. *J. of the Engineering Mechanics Division, Proc, ASCE,* Vol.108, EM5, 1982, 764-782.

Bicanic N. and Zienkiewicz O. C. constitutive model for concrete under dynamic loading. *Earthquake engineering and structural dynamics,* 1983, 11: 689-710.

Bischoff P H, Perry S H. Compressive behavior of concrete at high strain rates. *Materials and Structures,* V. 24, 1991, pp. 425-450.

Bischoff P. H., Bachmann H and Eibl J. Microcrack development during high strain loading of concrete in compression. Proc. *Euro. Conf. on Strut. Dyn.,* Eurodyna'90, Bochum, Balkema, Rotterdam, The Netherlands.

Cadoni E., Labibes K., Albertini C., Berra M. and Giangrasso M. Strain-rate effect on the tensile behaviour of concrete at different relative humidity levels. *Materials and Structures*, 2001, 34: 21-26.

Candappa D C, Sanjayan J G and Setunge S. Complete triaxial stress-strain curves of high-strength concrete. *Journal of Materials in Civil Engineering,* 2001, V. 13, No.3, pp. 209-215.

Candappa D C, Setunge S and Sanjayan J G. Stress versus strain relationship of high strength concrete under high lateral confinement. *Cement and Concrete Research,* V. 29, No.12, 1999, pp. 1977-1982

CEB-FIP Model Code 1990: design code/Comité Euro-International du Béton. Wiltshire: *Redwood Books,* 1993.

Comborde F, Mariotti C. Etude du comportement dynamique des materiaux fragiles par les elements discrets. 14 eme Congres Francais de Mecanique-Toulouse'99, CD-ROM, 1999.

Comite Euro-International du Beton - Federation Internationale de la Precontrainte. CEB-FIP Model Code 90 Redwood Books, Trowbridge, Wiltshire, Great Britain, 1990.

Cowell W. L. Dynamic properties of plain Portland cement concrete. Technical Report R447, Naval Civil Engineering Laboratory, Port Hueneme, Calif., June 1966, 46.

Eibl J. and Schmidt-Hurtienne B. Strain-rate-sensitive constitutive law for concrete. *Journal of Engineering Mechanics*, 1999, 125(12): 1411-1420.

Elfahal M M, Krauthammer T, Ohno T, et al. Size effect for normal strength concrete cylinders subjected to axial impact. *International Journal of Impact Engineering,* 2005, 31: 461-481.

Elvery R. H. and Haroun W. A. A direct tensile test for concrete under long or short-term loading. *Magazine and Concrete Research*. 1968, 20 (63): 111–116.

Fujikake K, Mori K, et al. Dynamic properties of concrete materials with high rates of tri-axial compressive loads. *Structures and Materials,* v 8, Structures under Shock and Impact VI, 2000, pp. 511-522.

Gran J K, Florence A L, Colton J D. Dynamic triaxial tests of high strength concrete. *ASCE Journal of Engineering Mechanics,* V. 115, No.5, 1989, pp. 891-904.

Grotes D. L., Park S. W., Zhou M. Dynamic behavior of concrete at high strain rates and pressures: I. experimental characterization, *International Journal of Impact Engineering,* 2001, 25: 869-886.

Harsh S., Shen Z. and Darwin D. Strain-rate sensitive behavior of cement paste and mortar in compression. *ACI Materials Journal,* 1990, 87(5): 508-516.

Hsu T. T. C., Slate F. O., Sturman G. M. and Winter G. Microcracking of plain concrete and the shape of the stress-strain curve. *ACI Materials Journal,* 60(2) (1963) 209-224.

Imran I, Pantazopoulou S J. Experimental study of plain concrete under triaxial stress. *ACI Materials Journal,* V. 93, No.6, 1996, pp. 589-601.

Kaplan S. A. Factors affecting the relationship between rate of loading and measured compressive strength of concrete. *Magazine of Concrete Research,* 1980, 32(111): 79-88.

Kupfer H., Gerstle K. H. Behavior of concrete under biaxial stresses. *Journal of the Engineering Mechanics Division,* ASCE, August 1973, 99(EM4): 852~866.

Lee S. K., Song Y. C. and Han S.H. Biaxial behavior of plain concrete of nuclear containment building. *Nuclear Engineering and Design,* 2004, 2 (27): 143-153

Lin G., Yan D. and Yuan Y. Response of concrete to dynamic elevated-amplitude cyclic tension. *ACI Materials Journal,* 2007, 104(6): 561-566.

Li W and Xu J. Impact characterization of basalt fiber reinforced geopolymeric concrete using a 100-mm-diameter split Hopkinson pressure bar. *Materials Science and Engineering:* A 513–514 (2009) 145–153

Ma H., Chen H. and Li B. Influence of meso-structure heterogeneity on dynamic bending strength of concrete. *Journal of Hydraulic Engineering,* 2005, 36(7): 846-852. (In Chinese).

Malvar L J, Ross C A. Review of strain rate effects for concrete in tension. *ACI Materials Journal,* V. 95, No. 6, 1998, pp. 435-439.

Mellinger F. M., Birkimer D. L. Measurement of stress and strain on cylindrical test specimens of rock and concrete under impact loading. Technical Report 4-46. U.S.*Army Corps of Engineers, Ohio River Division Laboratories, Cincinnati.*Ohio. Apr. 1966, 71pp

Milashi H., Sasaki T. and Izumi M. Failure process of concrete: crack initiation and propagation. ICM 3, Cambridge, England, vol.3, 1979. p. 97-107.

Newman, J B. Concrete under complex stresses. Development in concrete technology-1, Lydon F D (Ed.), *Applied Science Pub,*London, 1979, pp.151-219.

Paulmann K. and Steinert J. Beton bei sehr kurzer Belastungsgeschichte (Concrete under very short-term loading), *Beton*, 1982, 32(6): 225-228.

Reinhardt H. W., Rossi P., van Mier J.G. M. Joint investigation of concrete at high rates of loading, *Materials and Structures,* 1990, 23(135): 213-216.

Richart F E, Brandtzaeg A. and Brown R L. A study of the failure of concrete under combined compressive stresses. Bulletin No.185, Engineering Experiment Station, University of Illinois, Urbana, 1928.

Ross C. A., Jerome D. M., Tedesco J. W. and Hughes M. L. Moisture and strain rate effects on concrete strength. *ACI Materials Journal,* 1996, 93(3): 293-300.

Ross C. A., Tedesco J. W. and Kuennen S. T. Effects of strain rate on concrete strength, *ACI Materials Journal.*92 (1) (1995) 37–47.

Ross C. A., Thompson T. Y. and Tedesco J. W. Split-Hopkinson-pressure-bar tests on concrete in tension and compression. *ACI Materials Journal,* 1989, 86(5):475-481.

Rossi P. and Toutlemonde F. Effect of loading rate on the tensile behavior of concrete: description of the physical mechanisms. *Materials and Structures,* 1996, 29(186):116-118.

Rossi P., Van Mier J. G. N., etc. Effect of Loading Rate on the Strength of Concrete Subjected to Uniaxial Tension. *Materials and Structures,* 1994, 27:260-264.

Lee S. K., Song Y. C. and Han S. H. Biaxial behavior of plain concrete of nuclear containment building. *Nuclear Engineering and Design*, 2004, 2 (27): 143-153

Sercombe J., Ulm F. J. and Toutlemonde. Viscous hardening plasticity for concrete in high-rate dynamics. *Journal of Engineering Mechanics,ASCE,* 1998, 124(9): 1050-1057.

Sfer Domingo, Carol Ignacio, Gettu Ravindra, et al. Study of the behavior of concrete under triaxial compression. *Journal of Engineering Mechanics,* V.128, No.2, 2002, pp. 156-163.

Swaddiwudhipong S., Lu H. and Wee T. Direct tension test and tensile strain capacity of concrete at early age, *Cement and Concrete Research,* 33 (2003) 2077–2084

Takeda J, Hiroyuki T, et al. Mechanical behavior of concrete under higher loading rate than in static test. *Journal of the Institution of Water Engineers and Scientists,* 1974, pp. 479-486.

Takeda J. and Tachikawa H. The mechanical properties of several kinds of concrete at compressive, tensile, and flexural tests in high rates of loading, *Trans Architect. Inst. Jpn,* No. 77(1962)1-6. (In Japanese).

Tedesco J. W., Ross C. A. and Kuennen S. T. Experimental and numerical analysis of high strain rate splitting tensile tests. *ACI Materials Journal,* 1993, 90(2): 162-169.

Xie N.X. and Liu W.Y., Determining tensile properties of mass concrete by direct tensile test, *ACI Material Journal.* 1989, 86 (3): 214– 219.

Yan D. and Lin G. Dynamic properties of concrete in direct tension. *Cement and Concrete Research,* 36(2006)1371-1378.

Yan D. Lin G. and Chen G. Dynamic properties of plain concrete in triaxial stress states. *ACI Materials Journal.* 2009, 106(1): 89-94.

Yan D. and Lin G. Influence of initial static stress on dynamic properties of concrete, *Cement and Concrete Composites,* 2008, 30:327-333.

Yan D. and Lin G. Dynamic behavior of concrete in biaxial compression. *Magazine of Concrete Research,* 2007, 59(1):45-52.

Zheng D., Li Q. and Wang L. Rate effect of concrete strength under initial static loading. *Engineering Fracture Mechanics,* 74(2007), 2311-2319.

Zielinski A. J., Reinhardt H. W. and Kormeling H. A. Experiments on concrete under uniaxial impact tensile loading. *Materials and Constructions,* 1981, 14:103-112.

INDEX